MAKING MONEY IN FILM AND VIDEO

SECOND EDITION

MAKING MONEY IN FILM AND VIDEO

SECOND EDITION

A Freelancer's Handbook

Raul daSilva

Focal Press
Boston London

Focal Press is an imprint of Butterworth–Heinemann.

Cover art and illustrations for Chapters 3, 4, and 7 by Ray daSilva. All other illustrations by Sylvain Despretz.

Recognizing the importance of preserving what has been written, it is the policy of Butterworth–Heinemann to have the books it publishes printed on acid-free paper, and we exert our best efforts to that end.

Library of Congress Cataloging in Publication Data

daSilva, Raul.
Making money in film and video : a freelancer's handbook / by Raul daSilva. —2nd ed.
p. cm.
Includes bibliographical references and index.
ISBN 0-240-80144-X (pbk. : alk. paper)
1. Motion pictures—Production and direction—Vocational guidance.
I. Title.
PN1995.9.P7D3 1992
791.43'0232'023—dc20 92-13908

British Library Cataloguing-in-Publication Data

A catalogue record for this book is available from the British Library.

Butterworth–Heinemann
80 Montvale Avenue
Stoneham, MA 02180

10 9 8 7 6 5 4 3 2 1

Printed in the United States of America

To
Jamison Handy

CONTENTS

PROLOGUE

Max Cleland, who was left with only one arm and no legs by the Vietnam experience, was good enough to send me a poem I heard him read when he took office as the head of the Veterans Administration in the mid-1970s. The poem was written by an unknown Confederate soldier.

Those of you who are at the midway point or beyond in your careers in what must be the cruelest of all possible professions will understand the appropriateness of my now passing on to you this poem.

Prayer

I asked God for strength, that I might achieve,
I was made weak, that I might learn humbly to obey.
I asked for health, that I might do greater things,
I was given infirmity, that I might do better things.
I asked for riches, that I might be happy,
I was given poverty, that I might be wise.
I asked for power, that I might have the praise of men,
I was given weakness, that I might feel the need of God.
I asked for all things, that I might enjoy life,
I was given life, that I might enjoy all things.
I got nothing that I asked for—but everything I had hoped for.
Almost despite myself, my unspoken prayers were answered.
I am, among all men, most richly blessed.

PREFACE

During the 1970s and 1980s, film schools and film departments proliferated throughout the United States. The core of their curricula was courses in production techniques and aesthetics. Most were taught by academicians thoroughly schooled in the theory of filmmaking but generally without practical experience in the field. As a result, they were unable to provide their students with insight into the daily task of extracting a livelihood from the film industry.

Film school graduates began to discover that, whereas they may have been equipped to make a film, they were unqualified to sell their expertise. They had little or no idea where to take their skills. Who needs film? When? Where? For what reasons? How much is charged for time? How much, if anything, should they tack on to the purchase price of items needed to complete a project? These and hundreds of other questions had to be answered, and they could be answered only by an experienced filmmaker.*

The plight of these film-school graduates became apparent to me as I interviewed some of them for jobs in my capacity as an executive producer at an advertising agency. Not only did they have no training in seeking work, they also had no idea how to write a simple film proposal. They were ignorant of corporate structure and lacked knowledge of advertising, business communications, public relations practices, sales promotion, and the marketing process. Each of these subjects has a practical importance for the working filmmaker.

*Note: Throughout this book, except where otherwise specified, the term *filmmaker* means anyone engaged in the enterprise and profession of creating screen entertainment and/or communications for television, cable, or video or for projection, as in a film theater.

As a first attempt to address this problem, I wrote a book in 1978 entitled *The Business of Filmmaking*. Although that book explored the intricacies of budgeting and writing proposals, it was a simple introduction to the problem, the tip of the iceberg. Then in 1979 came Mollie Gregory's *Making Films Your Business*, which is a nice expansion on the subject of filmmaking business problems but is not a handbook on survival tactics in screened communications.

The book you hold in your hand is the first of its kind, a thorough exploration of the real world of the filmmaking business. I hope it will go a long way toward answering the many questions you will have as your career unfolds. It will not answer every question, but I guarantee that it will give you the tools necessary to find the answers yourself. On first reading, some of the sections—for example, that on advertising-agency personnel—may seem superfluous, but that impression will die quickly once you actually begin to grapple with the daily problems of seeking assignments from advertising shops and large corporations.

The purpose of this book is to build a bridge from the theoretical world of academia to the practical world and hard-nosed reality of filmmaking. It is not a production or technical manual of any kind. Rather, it presumes certain basic technical knowledge on the part of the reader and, therefore, uses the terms and phrases of filmmaking vernacular to achieve understanding. The Glossary includes the more important of these terms and others that help the filmmaker, but it does not provide the extensive and constantly changing vocabulary that a production-technology book might. It covers instead the terms of selling and marketing that form a basic, unchanging language decades old.

This is a lean, working handbook that should go a long way toward easing you into the business. It is a reference book and has been written to sit on your desk for repeated review until the experience contained within its covers becomes your own. Readers who have already been out in the field may recognize some of the problems described and perhaps may have discovered a new solution to them. In some cases, it may contain the overlooked piece to an unsolved business puzzle.

Making Money in Film and Video was written to help you survive and succeed in a most competitive business. I have been in it long enough to understand that such competition is good, that your success will also help mine. All of life works that way. We are all interconnected in an enterprise that is both a cause and a way of life. My success is also yours. Go out and make a good film.

1
INTO THE TWENTY-FIRST CENTURY

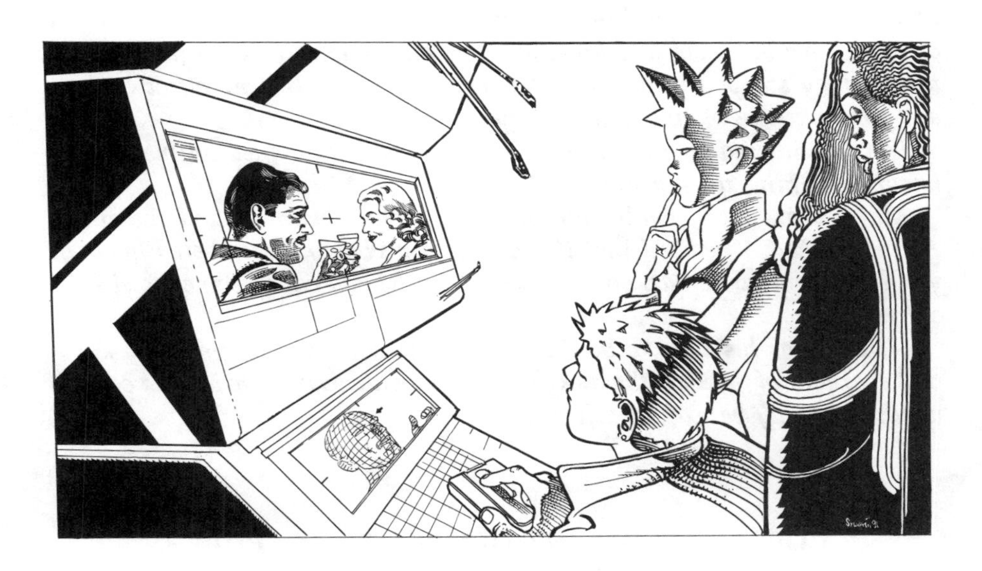

Sometimes everyone else's life appears to be better than one's own. In my case, my family seemed to have moved around a great deal while I was growing up, and I missed the hometown life I saw depicted in movies of Americana during my boyhood. Later, I realized that while I did not have the comfort of the familiar, I had the richness and excitement of the new.

We lived in Astoria, Queens, New York, for a few years not far from where Chester Carlson, while meditating one day in 1939, envisioned the electrostatic process that was to become the impetus behind Haloid, a small Rochester, New York-based company, that later came to be known as the Xerox Corporation.

Also, by odd coincidence, I lived a few blocks from the Astoria Studios, where motion pictures began long before my childhood and before Hollywood was anything but a tiny hamlet surrounded by orange groves.

In those days of my youth, there were still a lot of stately mansions on several streets in Astoria, and one of them was owned by a man named John Chalmers. Mr. Chalmers was not a popular man in the neighborhood, as he constantly chased the children away from his front lawn, explaining it was not a public park. Oddly enough, although I was one of the kids he chased, I understood his position and one day knocked on his door to ask him if he needed someone to mow the large expanse of grass. He gave me the job and, for the rest of my family's stay in Astoria, I had a tiny income.

"You old coot," the kids used to shout. Mr. Chalmers would always smile back, unperturbed. "You old this and that," the kids continued, realizing they were not getting to him. Sometimes Mr. Chalmers would chuckle a bit and wave back, smiling broadly. They all thought he was crazy. Eventually, they stopped calling him names, realizing it had no effect on him and was a waste of time and energy.

But I wondered about old Mr. Chalmers, and it dawned on me that he knew something we did not know.

Many years later, while ruminating on John Chalmers and the entire Astoria period, I realized that the old timer probably remembered his own youth and, in particular, one fine day when he had called someone else an old so-and-so.

Now it was his turn to be the "old so-and-so." And he probably knew that everyone of the kids, the lucky ones who would survive to be "old coots," would someday be exactly where he was.

A DREAMER, A FEW STILL PHOTOGRAPHERS, AND A PSYCHIATRIST

Consider that it was only in the early 1890s that motion picture technology actually came together. The famous former student of Freud, psychologist and psychiatrist Carl Gustav Jung, discovered and labeled a phenomenon that he called *synchronicity* whereby concepts and ideas (events) appear simultaneously in different parts of the world, emerging from the timeless unconscious. The motion picture is certainly an example of synchronicity.

In 1894, while seeing Edison's Kinetoscope demonstrated in Paris, Louis Lumiere became convinced that a similar series of sequential pictures could also be put on film and projected onto a large screen. In 1889, George Eastman's engineers, working feverishly in Rochester, New York, had invented the sprocket hole, which revolutionized motion picture film, enabling W.K.L. Dickson and Thomas Alva Edison to perfect their motion picture projector and to begin production in 35mm film just a few years later.

Nearly ten years earlier, however, England's William Friese-Greene had also worked on various motion picture projectors. None was successful but his work helped to inspire the birth and growth of the industry.

Now just ten decades later, in the early 1990s, digital electronic recording is rapidly replacing organic film stock, which will soon disappear forever. Not many years ago, all film editors worked by physically manipulating film stock, cutting and splicing it. Today, much of the editing is done on a PC keyboard, as many editors work by viewing scenes and keying their placement instructions into a computer.

With all this social and technological change, is human nature also changing? A simple examination of a few of William Shakespeare's 400-year-old plays shows that, for good or ill, the human personality seems to remain the same. While the frame of reference is in a constant state of change, the human personality, apparently, is a constant.

THE FILMMAKER AT THE TURN OF THE CENTURY

With the understanding of the constancy of human behavior and the amazingly rapid acceleration of technology, you can arm yourself for survival. One way to arm is to understand the socio-political landscape and the texture and nuance of interpersonal relationships. For now, you would do well to consider the nature of film and video.

In the following chapters, you will be shown how human nature, technology, and the challenge of surviving in a competitive industry effect how you look for and accomplish freelance jobs.

Chapter 2 covers the importance of adequate preparation, a good point to remember when seriously interested in success. After that will be covered the various problems most people face in the pursuit of a career.

Chapter 3 looks at how you solicit new business, especially if you have no prior experience. Chapters 4 through 7 explore the various techniques and skills vital for continued success in this field, such as dealing with clients, public relations, and professionalism in the field. Chapters 8, 9, and 10 look into areas where potential exists for growth, and 11 and 12 highlight the area of greatest potential, corporations. Chapter 13 looks at the elements of success.

End-of-Millenium Opportunities for the Filmmaker

It is not too early to begin positioning yourself for a role in the coming celebration of the great historical event—the end of the millenium. Opportunities will abound for communicators and entertainers in screen media. The opening up of the Eastern Bloc and the democratization and fragmentation of the old Soviet Union will usher in opportunities for those interested in international work. There will be opportunities found throughout the spectrum of film, video, television, and cable programming both in the informational and entertainment sectors.

Cable TV will be a major force, while home-video entertainment, still in its infancy, has already bypassed theatrical presentations. Also clearly seen, advanced forms of digital recording on objects as small as a credit card are on the horizon, as has been seen for over a quarter of a century on "Star Trek." But, of course, the format does not matter to the filmmaker.

While realizing this important moment in history exists, do not wait to be inspired and motivated by your external world. Create it for yourself from within. If you're not now a self-starter, learn techniques to keep yourself motivated. Your competition will most certainly be doing that.

REVIEWING THE 1980s AS PREPARATION FOR THE 1990s AND THE NEW CENTURY

As mentioned before, keen awareness of your socioeconomical milieu is critical. During the 1980s when money was far cheaper and more fluid, it was used to "light cigars," much as was done in the 1920s before the great crash of '29. Now in the 1990s, with the pendulum swinging back, it is clear that this decade will not be known as the "Gay Nineties," as was the case in the last century.

For the filmmaker, all time is expensive, not just overtime and not just the moment when the sun is at a special position on the horizon and its radiation is beautifully screened as it cuts through the late-day atmosphere.

Everyone knows that the best directors are those who have prepared themselves before coming on the set. They not only have worked out the characterizations well but have intimate knowledge of the conflicts involved and how each character functions to meet conflict head on with variations caused by differing personalities. The good director has also rehearsed the actors well and readied them for that moment when tape or film stock is flowing through the camera recording head or film gate, being exposed to special light in an environment carefully selected or created specifically for the shot—when all time is golden, especially with today's tightening budgets.

Realities in the Marketplace

If you have stars in your eyes and believe you have a good chance to make it in Hollywood, think again. The odds are not good. Logically, the numbers do not make sense. Targeting a career in Hollywood filmmaking is not only high risk, but the mythical ego rewards are simply not there. Even if by some miracle of dumb luck (talent is a given) you happen to sell a scenario or get a call to direct a film, it might very well be your only one. The opportunities for creative control do not exist for the most part, especially for writers. Directors go to bat and are immediately confronted with barriers to their creativity from within and without the production.

Also, if you are interested in the "great wealth" you hear about in the media, you will soon come to realize, after the slightest bit of study, that your chances for getting rich in the *entertainment* sector are about the same as winning a state lottery or "breaking the bank" in Vegas or Atlantic City, that is, nil. Why be disappointed and waste a lot of time out of your life?

At this writing, theatrical, entertainment film-production budgets have reached obscene levels, yielding ultra-high risk and little or no returns to investors. Imagine how many pictures like the classics *The Best Years of Our Lives, Twelve Angry Men, How Green Was My Valley, The Quiet Man, Shane,* and *Citizen Kane* you could make with the budget of a single *Terminator* 2? The answer is no less than 50 or so, even with current pricing. Consider all the employment, the spread of wealth, the great potential of cultural gifts to society that were stolen by greed on that single production. *T2* is nothing but raw, repetitious violence with a story line so thin you cannot even bite into it. One leaves the theater feeling like a mugging victim, which is actually the case.

Notice that all the classic films have one thing in common, which *T2* lacked. What was that? If you guessed *good writing,* you guessed correctly. This means that the classics, great dramas about the victory of the human heart over adversity, will still be around a very long time after *T2* has been relegated to special-effects oblivion.

Because the entertainment business is such a miniscule part of the total filmmaking industry, most of this book will be devoted to informational production, where the true creative opportunities and money (albeit, no great wealth) really are. However, Chapter 10 will be devoted to giving as honest an analysis as possible of Hollywood.

How Technology Is Opening Up Opportunities for You

At this writing, Consumer's Union, a venerable and much imitated quality analysis group, came out with a report on television. A lot of what was predicted in the first edition of this book has become a reality. According to some 200,000 responses to a 1990 questionnaire addressed to the membership of Consumer's Union, the highest rated TV channels are informational channels. The Discovery Channel, which is all information, rated highest in this survey. Following closely are channels such as Cable News Network and PBS. At the bottom were the once all-powerful (only because they had a monopoly) national networks, ABC, CBS, and NBC. Almost as low in ratings is the new Fox Network with its barrage of yellow/tabloid journalism.

The horizons to cast your weary eyes upon are cable TV and home video. In cable TV, the technology is expanding to the point where many hundreds of new specialized channels will be opening as the old, thick, copper, low-capacity coaxial cable is replaced by a much thinner, lighter, optical-fiber cable capable of carrying many channels per strand through the use of laser light fed in ASCII code.

Digital compression will additionally multiply the new capability by four to eight times. A 150-channel system, for example, will go to 600 or 1200 channels.

At this juncture, the dark cable TV cloud on the horizon is that out of the great tangle of early cable companies, seven giant cable operators have emerged, companies such as Time Warner, Viacom, and Tele-Communications, Inc. Since they also own and control the product, the fear is that these giant companies will begin to stifle creativity as did the commercial networks when they ran the game. Even worse, each cable operator plays a market or territory completely unopposed by competition. It will be interesting to see how the entire industry matures, or whether, like commercial broadcast television, it will be gradually replaced by something that is not now in existence and that cannot now be foreseen. My bet would be on that possibility.

Summarizing this chapter, it can now be clearly seen that the only constant in the universe is change. While the human personality remains the same, with the age-old battle over fear, life and society are now on a fast track of technology and communications. The globe is now a giant electromagnetic grid. The world is wired; instant communications is now an absolute reality. The good student of this fast-track society will win.

Life itself is as a child. It grows in accord with the thoughts, acts, and overall behavior of society, acting exactly like a mirror and reflecting back its personality. No other time in all of recorded history has offered so much opportunity for so many to prevail.

2

BEFORE YOU GET STARTED

It was the middle of August and I was sitting straight in my chair to dry the perspiration off my back after walking eight midsummer Manhattan blocks from a mixing studio. The air conditioner was browning out, but I cared little because I was in the company of gods. The place was TeleCraft, over Sardi's, where many great animation spots have been cut. It is not just another editing studio but one of the world's best, a reputation earned by 35 years of quality and reliable work by editors responsible for some of the finest footage to have seen light.

That day a nicely dressed young man came in and looked toward the five of us—all over 30 and representing nearly 125 years' experience in films—sitting clustered in worn director's chairs. He wore a dark suit in the ungodly heat and a wet, white shirt with a grayish tie that looked like it had been loosened and retightened 40 times that day. He was, naturally, nervous.

"Er, hi, I . . . was wondering if you could use help, that is, a production assistant, editor, grip, or. . . ." His voice seemed to trail off. We exchanged

glances. A visiting producer from Rochester, New York, I declined to answer him. A more outspoken member of our group could not bear seeing the chap suffer anymore. Waving his hand around our circle of experienced filmmakers, he asked, "Do we look like we need any help?"

At a loss for an answer, the kid thanked us and left. I had a helpless tight feeling in my chest. I knew exactly how that young guy felt, even though I had not been in his position for well over 15 years.

GROUNDWORK

Filmmaking is one of the most difficult industries in the world to break into. It may offer great rewards, both spiritually and financially, but, the Hollywood story aside, the real business (and the one that offers the most openings) is film as communication, not entertainment. By communication film is meant educational and training films, audiovisual or presentations pictures, television spots, multimedia for trade shows, and, most importantly, videotape, a medium with which most filmmakers today will be involved at some point in their careers.

At last count, over 1,000 schools, colleges, and universities offered some kind of film, video, or audiovisual curriculum. Included in this number are the many small colleges with media departments that endeavor to give students some exposure to nonprint communications. Barely a handful of such programs existed two decades ago. What happened?

The population explosion happened. The need for information happened. The coming of age of young men and women who fought in wars where the impact of audiovisual propaganda and strategy was felt happened.

If you are reading this book, it is quite likely that you have completed a media curriculum and are ready to explore the moneymaking possibilities of film. This is the time to appraise the market and yourself; the first half of this chapter is designed to help you do just that. The second half, which covers advertising agency filmmakers, is offered to assist your self-evaluation. Dreams of an independent film empire aside, you may actually be better suited for agency work. Many independents first apprentice with agencies to acquire experience and something just as valuable—business contacts. At the very least, however, as an independent or freelance filmmaker, you should know how the other half lives.

Where the Action Is

The two largest areas of filmmaking, according to Tom Hope, a consultant who chronicles the audiovisual market each year with his *Hope Reports*, are business and education. These areas offer a broad range of possibilities for the filmmaker. Some, though far from all, are outlined in the following paragraphs, beginning in the business realms where audiovisual use is most prevalent.

The public relations offices of major corporations and civic organizations may turn to visuals to enliven corporate shareholders' meetings and annual reports; to present case histories and securities analyses; to enhance internal communications and community relations; or to boost product publicity, perhaps utilizing the corporate puff in a theatrical release.

Filmmaking techniques in advertising and sales promotion are more self-evident. Television commercials, short shorts for nontheatrical distribution, and merchandising films that enhance the product or service, perhaps at point of purchase, enjoy fairly large audiences, while films emphasizing sales presentation, new product demos, and sales training are targeted at smaller audiences.

Multiple uses of film and audiovisual materials can simplify management's work load while providing accurate and attention-grabbing information in personnel fields: technical, management, and skills training; labor relations; orientation; and behavioral education.

Beyond but not excluding the business world, there are also multiple educational markets for film. The medium can increase general awareness of the human condition by probing life values, establishing the need for mutual understanding, or exploring the impact of a larger population and an aging citizenship on the community. Humanity's relationship to its environment is explored in films on conservation, energy, city planning, transportation, and government.

Law enforcement agencies use film for training and informational purposes, while all medical disciplines from physiology and pathology to epidemiology and the study of social drug problems benefit from the filmmaker's ability to capture innovative techniques or historic breakthroughs and transmit them accurately to farflung audiences.

Finally, film can broaden the recreational experience. Ignoring (for the moment) Hollywood entertainment features, film enlarges sports experiences and can teach basic and even primitive arts.

In each of these areas, filmmakers are turning people away from the inefficiency and lack of immediacy of the print media and toward the sensory impact and practicality that film and other audiovisual media offer.

A Personal Inventory

Now you know the possibilities, but how do they apply to you? Can you answer the following questions without consulting a reference book: What is the function of a securities analyst? What is the function of a securities analyst's meeting and presentation? What exactly is community relations? Can you write a proposal for any of the business fields mentioned in the previous section? Can you write an effective budget estimate? Can you actually sell the need for film to a client? Unless you can answer yes to most of these questions, you have limited your scope in filmmaking and your spectrum of operations. You have some homework to do.

Notice that I did not ask you if you can write a script. You can always hire a writer. Or can you? Would you be able to edit a script for accuracy? Would you be able to evaluate it adequately? Or check it prior to presentation to a client to ascertain that all pertinent points are covered?

You say someone else would do that since you are a technician? It takes but a few weeks to learn how to operate a camera, read a light meter, check out the latest emulsions and hardware, read the lab standards book, and learn to operate an editing console or rewind bench. That is all, just a few weeks. I have seen it done; I have taken writers in hand and made producers out of them in a few weeks. But make writers out of producers? It cannot be done in a few weeks, seldom in months, and often not even in years.

Putting the writing aside, where do you fit if you lack a business background? Ask yourself if you could possibly help a producer who needs someone to produce and direct from hard script. Can you do that with the education you have? If so, you are on your way to point A.

Given the fact that you have acquired practical knowledge, you have arrived at point B in your career as a filmmaker—you have a fairly well-rounded background and understand exactly why films are made, not the artistic experimental film you made in school, but the films people buy. Can you progress to point C? Can you accept a world that has not fully made the transition from print to nonprint communications? You may understand the impressive efficiency and high-sensory impact of audiovisuals and films, but you must never forget that the average person views film solely as a form of entertainment. Even at this late date, typical businesspeople have a limited knowledge of film and its effectiveness. They have yet to be taught the medium. If you can educate them, you have reached point C.

Proceeding to point D, what do you know about distributing film? Are you completely familiar with organizations such as Modern Talking Pictures? If you do not know what such companies do, how they operate, and how much they charge, you have more homework to do. One of the worst things a maker of informational films can do is "hit-and-run" a client. Simply leaving a can of film with a client does not complete the job. You must know how the client can put that roll of images to work, and you must share this strategy with the client and account for it in your original budget. If you do not help your clients with distribution (and generally they will need help), you are simply handing over a vehicle with no fuel in it. The film may end up on a shelf after only a few screenings, and you will have destroyed the possibility of repeat business for both yourself and other filmmakers.

It is true that editors are still learning their trades after 25 years. And directors continue to discover new techniques after three decades in the genre. In the film business, you never stop learning.

The Starting Edge

In my experience, most applicants for agency production jobs possess single or multiple degrees in film, broadcasting, or mass communications. Unfortunately, most candidates lack the two most essential qualifications for communications filmmaking: the ability to write a script and a solid business and marketing background.

A student seeking a film degree should seriously consider taking sufficient elective courses in writing and business to qualify for an MBA program. A person with a marketing background who can at least edit a script has a chance to land a job. If one can write a good script based on sound knowledge of marketing technology, a job is probably waiting somewhere. An associate once conceded that he would rather hire people with marketing writing backgrounds and teach them film technology than the reverse. I tend to agree. Business is an infinitely larger and more complex field than filmmaking. Agency filmmakers can always depend on outside resources for film technology but must do their own marketing.

If you are currently studying film technology or are in the field already but considering further study, the following courses will prove helpful:

advertising theory
advertising design
display design
creative writing
screenwriting
journalism
public relations
marketing
sales promotion
retailing and distribution
business law
copyright law and ethics
management information systems
market research
marketing motivation and persuasion
personnel administration
personnel training methods
sales training
cost accounting
data processing (principles and systems)
computer programming

No matter to what area of filmmaking you aspire, the courses just mentioned compose an excellent adjunct to your film background. It can often save you when all else fails.

In addition, remember what was stated earlier: business films are where the action is. The field is wide open and growing. It will continue to grow as students who are experienced in the use of audiovisual technology replace business communicators who were weaned on outdated textbooks. Film/video is as fresh as today. It is fast. It has impact, stored energy, and utilizes more sensory input. For these reasons, business will continue to turn to video, film, and audiovisual media.

THE AGENCY OPTION

Among today's nontheatrical filmmakers, the title *producer* is becoming as vague as the once crystal-clear New York skyline. It can mean anything from filmmaker to backer or jobber. In the advertising agency, it formerly designated the person who had the television spots made, but this is no longer always the case. More agencies are developing in-house capabilities as films, video, and audiovisual media replace print in marketing communications.

The Agency Media Producer

In some medium-sized agencies, a distinction is made between media (television) producers and nonmedia (direct communications) filmmakers. The latter generally do not produce commercials, although their films do occasionally surface on television as short subjects in a select pattern of theatrical distribution.

Television producers in the larger agencies tend to have their own pecking order with a personnel matrix as complex as that of any large corporation. With few exceptions, the smaller agencies offer much greater latitude for people who want to do their own thing—sort of. I say "sort of" because a client still must be satisfied, as well as a creative staff of copywriters and art directors, their supervisors, and the account people. These people can hassle the filmmaker or media producer, but they are not of the same ilk. Each has a specific language, as does the filmmaker.

The large-agency television producer is responsible for commercials and usually has them made in production studios. Although the cash rewards in this field can be great, there will seldom be much personal creative satisfaction. It is a job best suited to the filmmaker who has an administrative, as opposed to a creative, orientation. Nevertheless, while not required to be personally creative, the administrative producer must know and understand how to deal with the creative personality and mentality. If agency work interests you, be sure you can evaluate yourself on this basis.

Television producers working for small agencies will generally work on smaller accounts with respectively smaller budgets than their large-agency counterparts. Spot producers working for small- to medium-sized agencies may do their own shooting and editing and will certainly handle their products to a greater degree than their big-time colleagues. The local-spot producer is even more involved in the technical work, sometimes from the assigning of storyboards through the final viewing and approval of answer prints. The small-agency producer will also handle radio spots, whereas such work is departmentalized in a larger shop. The trade-off is less money and pressure for greater individual expression and a more relaxed atmosphere. Nothing is being implied here; there are filmmakers suited to both situations.

Direct Communications Filmmakers

The second field open to agency filmmakers, nonmedia or direct communications filmmaking is growing slowly but steadily. Today, this filmmaker is more often called *in-house* or *corporate* producer. Here, filmmakers can often work almost alone if they are good enough, and, unlike their distant cousin, the in-house filmmaker operating from the basement of General Bombastics, Inc., they will have an adequate budget. The field also offers many rewards to the business filmmaker. Production companies like Jam Handy, Wilding, and MPO, which pioneered communications filmmaking, have been the training grounds for some of today's established filmmakers.

A trend has been established today for agencies to offer more to a client. In the past, it was enough simply to place a certain amount of advertising in strategic media. Now, companies are rediscovering that advertising campaigns must be reinforced by solid sales promotion and merchandising. Such programs are best produced by a single creative group to avoid needless duplication and to maintain campaign continuity with an integrated marketing strategy. Although the logic of this backup approach is understood, too often the campaign flounders at the marketing level. A beautiful print ad that stimulates interest is useless without further literature on the product and a sales representative available for quick follow-up.

The agency that houses a resident communications filmmaker will also offer clients across-the-board marketing communications services. This will be neither a "boutique" agency nor a creative consumer house handling cosmetics, food, and cigarette accounts. Only a few of the large agencies have developed these true collaborative or direct communications divisions. A fertile ground for the marketing communications filmmaker, this field offers the greatest opportunities for students now churning out their great school epic.

Film plays a wide and varied role in the marketing mix. Films are used in public service and other public relations, sales promotions and meetings, trade shows, shareholders' meetings, personnel training and orientation, retailing at point of sale, and more. Shot in 16mm or videotape for one purpose, the marketing film often is sent to the reductions lab for conversion to Super 8 continuous-loop cartridge for rear-screen display or to 8mm video for meetings or point of purchase.

In short, the agency nonmedia filmmaker will often handle the full gamut of multimedia production: films, audiovisual communications, disc recordings, slide shows, cassettes, filmstrips with and without sound, and videotape. It is a virtual tour de force for the creative visualizer.

The Agency Film: A Scenario

By now, the multiple levels at which the agency producer operates should be apparent, but I offer the following case history of production of a sponsored film, taken from my own experience, to bring home the point. For emphasis, the producer in this saga is you.

The account executive, after consultation with his account ad manager, calls and asks for a meeting. His client has a problem that might be solved with a visual presentation. In any event, he has run out of ideas and needs a specialist.

The meeting is held, but the account executive lacks facts that you need to evaluate the problem. You decide to meet the client for the first time over lunch. Although nothing substantial is accomplished at this first client meeting, you now know the client a little better and can size him up for approach and handling. You have also monitored his reaction to you. (And he owes you a lunch.)

You set the second meeting for early morning over coffee so you will both be alert. At this meeting, you discover that your client is licensed to sell a foreign-made vibrating pile driver that is ten times faster than the steam hammer and is virtually noiseless in driving caissons 50 feet into the ground. Your client has been flying prospects around the world to display this miraculous machine at work. His problem is how to get more exposure to his market. The solution seems very obvious to you because you have a cinematic orientation. The client does not.

You propose a short film that shows four different locations and applications of the machine, or different models of the machine, at work. He wants to know how much it will cost.

Although you will need to quote a ballpark figure, the filmmaker who tosses off figures based on so much per foot is always making a mistake. So you follow up your quote by offering to prepare a proposal based on your approach to the problem. The client agrees.

In the proposal, you write a rationale for the film, indicating how the product will be displayed in the film and how the client will benefit by using the film. You also include a distribution pattern.

When the client reads the proposal, he tells you he will never get this budget approved. Having anticipated that very problem, you zap him with a study comparing the cost of squiring prospects around the globe during one year with the cost of the film and its distribution.

The client is amazed. The film will actually cost less than 20 percent of last year's product sales presentation budget. He has to fight tooth and nail with established forces within the company, but he gets approval for the film.

Meanwhile, you realize that you can reduce the film to Super 8 and put it into a salesperson's projector for automatic showing on a continuous-loop cartridge. You also determine that the client's product is indeed news, so you offer the client 20 one-minute prints of a silent newspeg coupled with a one-minute news writeup that you will distribute yourself to television news directors in the client's choice of markets. You realize you can make this from your outtakes. The client buys it for an additional $3500.

You now write a script for the film that will be narrated over the footage and a second one-minute script for the television newspeg. When the script is accepted, you get final budget approval on the film and produce it.

The film produces under good conditions, and you show your interlock on deadline. Some minor changes are necessary, but you are still on the line because you budgeted carefully, allowing some margin for changes.

You are now faced with the arduous task of training the salespeople to operate a Super 8 projector or portable videocassette recorder. You carefully select one that is widely available and has a low breakdown history. You also arrange for continuing service on the projectors with a reputable equipment supplier.

You have now been through the projector so many times you feel you invented the machine, but your demonstration is a success and you have also made a few friends on the side. Although the program is an outstanding success, you never hear from the account executive;

but the client writes you a nice letter that will be useful on your next approach.

Experienced filmmakers will smile and say, "If only they were all that easy." No, they are not all that easy. You may have a beautiful film, but unless it helps the client, you have failed. The better sales person you are, the better writer and researcher, and the more information you have about your client, the greater your chance for success. The formula is the same whether you are an agency producer or an independent filmmaker: preparation plus work equal success.

3
STARTING UP

The day I met Collin was great only for Eskimo dogs; the city was buried under snow. Collin was looking for a job. I was an executive producer at a medium-sized advertising agency.

Collin was late for the interview. Looking out my window, I assumed it was the weather. The great elm tree that so often brought me peace and relief from job pressure was nothing but a vague shadow in a field of snowflakes. Momentarily, my thoughts drifted to California and orange groves. Then, the phone rang: Collin had arrived.

Collin was a blue-eyed, carrot-top kid, rangy with a bouncing gait that made his curly red hair resemble tangerine gelatin on a spring. He crossed the office in two steps and sat down before being invited to do so; he seemed to be in motion even while sitting.

Collin wanted a job in advertising, he told me without offering a greeting. He had been a film/video major in college and had taken one or two advertising courses. Interesting. His resume told me his only experience had been placing some ads as the manager of a local Burger King.

We chatted a bit and I said, "Tell me precisely the type of work in which you are interested." Collin's quick response was that he "would make a good spot producer." He had a sample reel of spots he had made in college. The reel, he claimed, smiling broadly, "contained some of the best ads, better than those you see on television." Collin had a nice smile: a good thing to have when you do not know what you are talking about.

As it turned out, Collin knew very little about how an ad was produced. He knew next to nothing about the local freelance talent and production shops that agencies rely upon for television broadcast. He was ignorant of the inner workings of an advertising agency. I told Collin all this and added that his picture of the advertising business was as clear as my vision of the elm outside my window, still veiled with snow.

Because I have never understood those who do not take the time to tell, I told Collin that his unpreparedness rendered the interview a general waste of time. "There are too many others who have done all their homework before they come for an interview," I said. Attempting to soften this jolt to his confidence, I suggested that he consider seeking work a serious game—a game with rules. In games, you are competing with others, so you figure out what you have to do to compete well and then you prepare. The same basic tactic must be used in the job game.

HANGING OUT YOUR SHINGLE

Some of you may have apprenticed to accomplished and established filmmakers; others may have graduated from a film school or have earned an academic degree in filmmaking. Whatever your experience, you are currently frustrated for one of two reasons: either you cannot land a job anywhere, despite your willingness to sweep the floors for free the first year, or you already have a job but would prefer to work on your own for reasons related to individual style or life objectives. In either case, you have decided to make the big leap into independent production. But how do you get started in the world of filmmaking or audiovisual production?

Establishing Credentials

The first step in getting your business started is to establish yourself as a business entity. This can be accomplished by incorporation, which offers certain tax advantages and legal protection from libel, personal bankruptcy, and negligence. For most one- or two-person production companies planning to use freelance crews, however, incorporation will be a waste of time and money. Instead, you can set up in business by filing a DBA (Doing Business As) form with your county clerk. This process generally requires nothing more than recording the name of your business with the county. You can use your own name, but I would recommend using a catchy name with a film theme until you become better known.

If you think incorporation may be to your advantage, by all means investigate your options. A corporate lawyer and accountant will be able to provide invaluable advice (although you do not necessarily need an attorney to incorporate). Their combined knowledge and your own professional goals should permit you to judge whether incorporation is appropriate.

After selecting a name and registering as a business, your immediate needs are a phone listing, business cards, and stationery. Get the first by calling the business office of the local telephone company. Explain your situation to the service representative and request information on telephone company business accounts, various types of listings, and rates for Yellow Pages listing. Also ask for information on listing services in nearby cities or communities.

When your telephone is installed, have a printer make up 500 business cards that include your name, address, and telephone number. If you can afford it, have a local art studio design an eye-catching logo or select a distinctive typeface for your name. They will provide camera-ready art for the printer and can also prepare mechanicals for letterhead stationery, envelopes, invoice forms, and notepaper using the same design. Initially, I suggest you order only business cards, letterhead, and envelopes from the printer. You will need stationery to solicit work by mail and answer correspondence; for billing, simply type INVOICE across the top of the letterhead. The time to print special invoice forms is after your business is established. It is always wise to keep costs at an absolute minimum in the beginning.

Determining the Direction of Your Business

It takes an organized individual to succeed in business, and organization extends to determining the direction a business will take. Now that you have your business identity and rudimentary supplies, your next priority is deciding what kinds of films you want to make and for what group of clients.

This first decision is an important one, so take your time and select one area of filmmaking to pursue. Go with your strengths. A multimedia specialist, for example, has a brand of creativity quite different from that of the advertising-spot director. The former is experienced with a wide range of electronic hardware; the latter is skilled at making the most of tight time constraints. Although these areas were once handled by the same group of people, they are now so complex that they have become exclusive of one another. Specialization is the norm; therefore, it is essential that you choose a single field and work at it diligently until you become known in that field for your outstanding work. After you are thus established, you can then expand into another area.

DEVELOPING A BUSINESS PLAN

If you do not know about business plans, go to the library and ask the business- (or general-) information person about a book to help you write one. There is also software available to help in the writing of a business plan. This is extremely important when setting up a business of any kind.

The typical business plan has the following contents:

- rationale or description of the business, including detail on product and/or services to be offered and to whom
- principals involved with background profiles
- a marketing plan, which includes how the sales and sales promotion, the advertising, and possibly public relations will operate
- amount of operating budget needed and how it will be applied (including *overhead,* that is, all costs of operation, including rent, insurance, salaries, taxes, etc.)
- projected income, both gross and net, including three cash-flow spreadsheets: 1) worst-case scenario; 2) average case; 3) best-case scenario

A business plan is like a blueprint or a road map. Its importance cannot be overstated. If you start out without one, you are diminishing your chances for success.

SELECTING EQUIPMENT

Now the adventure really begins—you must invest some capital in equipment. You must know what type of camera you will need—three or four formats of video, and which film, 16mm or 35mm?

Remember, I said early on that clients will not be impressed unless you have your own gear. You must also be absolutely sure you buy the "in" equipment, especially if you are in a larger city, where the pretension often rises to the point of hysteria. For this, you must know other filmmakers in the area where you are planning to set up shop. There is just no other way of knowing the "inside" stuff because it is a social phenomenon. Strong personalities who are absolutely filled to the brim with self-confidence can acquire anything that works. For most, I would recommend going with the flow.

Obtain the very basic equipment first and expand according to your needs and the type of assignments you will be finding in the field. If you are planning on internal communications and training offerings to clients, you would be best with video. If you are approaching a public relations department and know that your audiences are to be found in theaters or auditoriums (large-screen projection), then move into film. At this writing (it changes fast), the only use for 35mm equipment outside of Hollywood is for large-budget commercials. Using 16mm equipment is fine for small-budget commercials that can be transferred to tape for posting to end up with a "film look."

Intimate knowledge of equipment is another key to your survival. Do not depend just upon the information you obtained in film or media school. In many cases, it will not match streetwise information.

Do not overkill. Use the minimum professional (not amateur) equipment needed. Even if you have a rich uncle who has extended his generosity to fund your studio, avoid going equipment happy. You will discover that, eventually, every dollar will have an important target.

Invoicing Your Jobs

You are in filmmaking to earn a living, so you must understand how to bill your clients before you approach them. Always get your business transactions in writing. Your client will probably require a contract, but if not, insist on one in a businesslike manner. Film equipment dealers carry contract forms that can be modified to fit your operation. You can also devise one yourself. Just be certain that your agreements are in writing and signed by both parties.

Two standard methods are used by filmmakers for billing sponsors or clients. One establishes four invoice periods: the first payment is due upon assignment of the project; the second after the script is approved by the client; the third at the interlock phase, or when the A&B rolls are approved with the soundmix; and the final payment upon approval of the final answer print or delivery of the first release print. The second and more commonly used method is billing in thirds: one-third due upon receipt of the go-ahead or purchase order; one-third upon approval of the interlock by the client; and the final third upon delivery of the answer print. In both cases, any client changes, such as modifications on the agreed script or interlock at the shoot or after, should be invoiced at above and beyond the established contract fee.

There are sound reasons for billing in thirds. The first payment provides you with advance money to hire subcontractors, purchase raw film stock, and rent equipment; the second payment is invoiced to acknowledge the client's approval of your work; the final payment not only approves the final answer print, but removes you from responsibility for print quality from that point on, unless you contract separately to deliver a definite number of prints or all future prints.

Distribution

Supplying a number of prints or videocassettes upon request is one part of distribution. Although distribution is not truly the filmmaker's area of expertise or operations—most turn to professional distributors—some filmmakers do accommodate their clients. Helping your clients beyond the production phase will indicate your interest in their welfare, thereby establishing a strong, trusting relationship and enhancing your reputation.

Distribution should always be covered in a separate contract unrelated to production. Additional income can be made from print orders; be certain to scout photo labs and duplication services carefully prior to ordering copies and get competitive bids for footage whenever possible. You should be able to suggest a sensible initial quantity to the client if your distribution plan is based on a review of the film's objective and intended audience. Despite the attraction of getting reduced cost per foot if large quantities are ordered, keep the print order within reasonable and studied numbers. Extra prints and cassettes sitting on shelves collecting dust are constant reminders to your client that someone made a costly error. Such mistakes harm you as well as other filmmakers.

THE SAMPLE REEL: YOUR FIRST JOB

Your business preparations are almost complete: you have established a direction, purchased equipment, and explored standard administrative procedures. You are ready to seek a client. First, you will need a sample reel of your work; few prospective clients will want to pay for an intangible product by an unknown.

If you have apprenticed with someone in the profession or worked in an agency, you probably have a small corpus of work from which to compile a sample reel. Be careful to keep your samples short whenever possible. Most clients have neither the time nor the patience to sit through a 30-minute plant-tour film, even an award-winning one. If you have something unusual that is lengthy but shows exceptional filmic skill, just pull out one or two of your best sequences. When you solicit spots, do not show anything but your own spots, and then only a few of your very best. Since the prospect will assume that you are displaying your finest work, be certain that you do.

If you are just starting out or are not exceptionally proud of your previous work, now is the time to consider producing a speculation reel. Some filmmakers develop a speculation spot using an imaginary product or service. Others produce a spot for a well-known product and send it to the advertising agency handling the product, hoping to impress them enough to land an account. Another option is to create a film on speculation for a prospective client and then try to sell it to the client if the film is well received. Each option represents a great financial risk on your part, since you will not be reimbursed if the client does not buy. At least you will have produced a sample film; if you sell it, you will also have a reference from your first client to show to your next prospect. Do not just sit there with your video gear (or rudimentary 16mm camera) wondering how to get off the ground. Make a commitment, and then adapt the following advice to your own situation.

Selecting a Market

Think *market*. An example of an obvious market for a filmmaker would be a dealer in large equipment. This dealer needs a film. In fact, he needs a film every month or two. Why? Because he has a huge customer list with varied needs. He currently depends on a handful of poorly photographed brochures; action motion picture footage would enable him to reach his market more effectively. Your job is to present this concept to him and to show him how a good film can save him time and fuel by making fewer product demonstrations necessary, thus increasing sales at lower cost.

Now that you have chosen a market, you must canvass the lucky company. Avoid asking your prospect whether he wants to buy a film; the answer will invariably be no. Instead, explain that you wish to use his company as an example on your sample reel. Why? Because your sample will provide other businesses in your area with the chance to evaluate a time-proven method of demonstrating products and services. You merely want to film his operation and show your prospective clients the effectiveness of demonstration on film and how it can save them time and money. Customers can be shown a film of several pieces of equipment in operation without ever having to leave the office. Remember, right now you only want his permission to film. After you have produced your sample, you can attempt to sell him the completed film.

Some Preproduction Tips

The impact of a professional voice-and-sound track will be beneficial in selling your film to prospective clients. Therefore, it is wise to invest in a quality voice. Do not go to a talent agency, but try approaching a good college actor or perhaps a voice you like on local radio or television. Such a person may be willing to do your film voiceover on spec (deferred payment when and if the film makes money) or for a low fee the first time out. (Voices start at around $50 an hour and climb—especially when agented—to over $2000.) The most successful producers generally select specialized talent for specialized jobs. Some scripts demand voices that convey humor, whereas others require authoritative deliveries. Take your time in selecting the right voice; the time will be well invested.

Once you have chosen the appropriate voice, have it professionally recorded. Talk to your local sound studio and develop a rapport with them. Nurture a long-term relationship and negotiate prices. Occasionally, a studio will throw in a quick mixdown for you if you have all the elements tucked under your arm when you take your talent into the studio for recording. So, be prepared. Keep costs low by recording short sequences of informal narrative with plenty of space between complete thoughts. This will enable you to bring up background music during the pauses. The mixed track can then be transferred directly onto the magnetic-stripe stock by connecting your tape recorder to your sound projector—if it has a sound-record mode. This method provides plenty of margin for alignment and decreases synchronization problems.

A sound caveat (excuse the pun): do not use commercial recorded music for background. The current sound record copyright law prohibits the copying of commercial sound recordings for this purpose. Producers must either contract for use of resources sold by a local audiovisual dealer or turn to commercial sound and music libraries. Such libraries are advertised in most trade magazines and provide the purchaser with original material released for low-cost use, music in the public domain, and standard sound effects. Building your own sound-effects library is a good long-term investment, and the cost need not be high.

Budgeting Your Film

Prior to beginning any film, whether it be a sample or a sponsor-contracted film, prepare a budget. Consider the value of your time, the cost of supplies and equipment, and assistants' fees. What should the budget be? Some producers charge as much as they can get away with, but that type seldom stays in business very long.

First and foremost, you are a businessperson. Your second role is that of filmmaker. Your profits might derive from a 15 to 20 percent commission on everything you buy and handle during the project (such as film stock, equipment, set) plus a set hourly rate; you may also negotiate a package deal based on your estimated cost of operation.

Average hourly rates vary depending on geographic location. Naturally, you cannot charge inflated New York prices if you do business in Lansing, Michigan. Therefore, you must set your rates for time and services based on three variables: what your competition is charging, the costs of goods and services, and how you rate yourself professionally. Are you just starting and learning on the job? Or will you be doing a fast and slick job, knowledgeably cutting corners while maintaining quality to save your client money? In the latter case, you should be earning more than the neophyte. Rate yourself honestly and sell your time in the area of $40 to $100 per hour.

Organize your budget using the sample budget log as a model.

Again, you will have to estimate what to charge and make personal allowances. At this point in your career you may choose not to charge for meetings and scouting, but on complex jobs where those

SAMPLE FILM PRODUCTION BUDGET

(nontheatrical or informational film or video)

1. Production expenses

a) Film stock or tape cassettes $ ________
b) Sound effects and music purchases $ ________
c) Sound studio rental $ ________
d) Talent (voiceover) $ ________
e) Sound mix $ ________
f) Processing of film stock $ ________
g) 20% commission of all purchase $ ________

2. Time (billed at hourly rate multiplied by time)

a) Meetings $ ________
b) Scouting locations $ ________
c) Script writing $ ________
(If you do not write the script yourself, but hire the writer, place this item under "Production Expenses.")
d) Camera and support crew $ ________

3. Sound studio direction $ ________

4. Editing, postproduction (titles, graphics) $ ________

5. General production management $ ________

6. Miscellaneous (shipping, faxes, etc.) $ ________

7. Insurance $ ________

Subtotal $ ________

8. 20% contingency* $ ________

Total $ ________

*A contingency is a reserve emergency fund, used in the event of a budget overrun. It is not good policy nor does it create a good producer's record to *plan* on using this reserve. Sponsors of films will review a producer's past fiscal containment carefully and generally react to a budget overrun negatively.

items eat up valuable time, you should itemize them. If you hire a scriptwriter instead of writing your own script, itemize this cost as a production expense; you are also entitled to a commission on it.

Bear in mind the following budget tips: when purchasing film stock, allow a five-to-one shooting ratio; if you are planning a five-minute film, purchase 25 minutes of stock. Sound effects and music can cost $100 per needle drop or more. This is where investment in a sound effects library can save money. If you do not have a mixing facility, it will cost $125 or more an hour for studio time. When you can do your own mixing and duplicating, you should itemize them under time costs.

Two places in your budget are appropriate for miscellaneous expenses. The first—clearly labeled miscellaneous—represents nonitemized production expenses such as travel mileage, tolls, and lunches. Keep receipts for these items. The second nonitemized entry is general production management time, under which falls all time spent on phone calls, making purchases, and handling materials. Although you can only estimate such items on your first budget, they represent valid costs that you will need to recoup. Your accuracy in anticipating these expenditures will improve with experience.

Insurance is definitely not an item to overlook. It is absolutely necessary that you insure against personal liability and against loss of production costs. Call local insurance agencies for a reference to a broker specializing in insuring filmmakers. To repeat, the contingency percentage (anywhere from 10 percent to 20 percent, depending upon your client's ability to acquire funds) is strictly to cover yourself in the event of an unexpected rise in expenditures caused by runaway inflation, a major accident, unusually bad weather, or any other unanticipated event and not expenses caused by negligence, poor administration, or bad management. Acquaint your client with this significant budget item. Also advise your client of the slight possibility that the contingency might rise above 20 percent in the event of calamitous unforeseen events.

This additional fund is not yours to keep if unused. If you have brought the production in under budget or at budget, the contingency fund returns to the client if it has been advanced to you. Most clients, however, will not advance it but hold it for its purpose: a contingency against emergency.

If this is your first film for profit, you may look at the bottom budget line and feel it is too high. Consider your client's budget and sales projections. If your film stimulates ten times its cost in additional business the first year—and perhaps more the next because of increased exposure—the initial expenditure will have been a wise investment indeed.

The Script

You have reached the final step before production: the script. Begin by drafting out a tight script outline that contains only factual information about the equipment and sets up the shots. Then, if you are a good writer, embroider your subject—but use restraint. If you are not and never will be a writer, locate and hire one. Check the Yellow Pages under writers or copywriters; ask local newspaper editors if they can recommend any freelance writers. Scout nearby colleges and night schools for instructors who teach scriptwriting. Do whatever is necessary to obtain a professionally written script.

There are several excellent books available that detail the mechanics of scriptwriting and offer annotated examples (see Bibliography). For our purposes, the following script of the opening of our sample film will get the cameras rolling:

SCENE 1: Establishing shot: Exterior of Agway equipment dealer's showroom. ZOOM SLOWLY FROM FRONT DOOR AND PAN TO ROW OF FARM IMPLEMENTS	
	Music up and falls behind narrator. NARRATOR: Central Agway has operated out of 2000 Elm Street since the days of Franklin Delano Roosevelt. Over the years, their reputation for service has become known throughout the county. (two-second pause)
2. SLOW PAN ACROSS LOT FROM HIGH POINT OF VIEW	
	Farm-equipment sales represent just one division of the large Elm Street Agway dealer. Let us look at some of the Agway products and services that can benefit you. (one-second pause.)

The script continues to display three or four products and services provided by the dealer. It need not be overloaded with words or effects. The combined impact of action film and sound will sell the products—and the film.

WHICH COMES FIRST, SCRIPT OR BUDGET?

In Hollywood feature films, you cannot hope to budget a production before first having a screenplay or scenario. This is primarily because you must abstract the "elements," or, as in object-oriented software, the parts of a screenplay (more appropriately called a *script* in the nontheatrical or informational production) that call for pricing.

Some of the elements of a feature screenplay are the various "talents," for example. Are they box-office? Are they unknowns? The difference can mean millions of dollars. Another element is the music score. Who does it—an unknown or someone who is "hot" or "in" for the moment?

Special effects is a third element. In *Terminator 2*, the special effects, mostly digital compositing and morphing (which have been around for years in nonfeature production and TV spots), suddenly became "hot," and film buffs took notice. The price, already inflated by the hot shop that did it (you know the one up in Marin County) because they were "in," was doubly inflated. You can avoid inflated costs if you wisely have the job done in an average city that might house the needed facility. All of these are elements that are abstracted from the screenplay to set a budget.

On the other hand, in informational films, you have a client. Generally, the client tells you, "I can afford $145,000 (or much less or more depending upon the client)." Then it is up to you to begin the task of creating a screen communication, all within a budget that includes everything from the price of the writer for hire, or freelancer (if you do not write it yourself or have a staff writer), to the price of film stock or videotape.

Making a film on budget is much like making a custom automobile or house. The buyer tells you how much is available and you create the product. It has always been so in custom production.

CAN A CLIENT AFFORD YOU?

This is an item rarely spoken about in film and video schools. In the New York City advertising industry, sometimes a client will actually be attracted by an "expensive" filmmaker. This immature and pretentious behavior is also irresponsible—it costs everyone because it drives up the price of goods and services and sometimes even puts people out of work.

It will best serve you and everyone else to avoid gouging and overpricing your work. It will also help to guarantee your survival as a filmmaker. Today's "hot" and pricey filmmaker is *always* tomorrow's cold potatoes. Also, once you set a high price, it is difficult to come down.

Production Steps

You are now ready to shoot. The following checklist should help you get your sample reel produced and ready for viewing.

1. If the film is not a pure documentary, scripting is done first. Otherwise, scout locations prior to scripting. Take meter readings at each site; make careful notes on angles.
2. Draft a good shooting script based on logical continuity and highlighting benefits.
3. Select crew and talent.
4. Film all sequences using the best possible exposures. Design your shooting schedule around efficiency, not continuity. Since film is almost always shot out of continuity, assign someone to follow continuity sharply. It helps to take instant stills to attach to the script.
5. For best results, have your film processed by a lab recommended by trusted associates.
6. Edit your film so that it makes visual sense.
7. Draft final narrative script from your shooting script for studio recording.
8. Obtain rough timing match—film tape—while recording narrative. Edit script further if necessary.
9. Prepare a good selection of light background music and appropriate sound effects.
10. Obtain studio soundmix, taking care not to disrupt the rough scene-to-scene synchronization of your final script (item 8 above).
11. Have an answer print made from your original film rolls at your local lab.

Sales Presentation

Your sample film is complete. Now you can attempt to sell it to your client/prospect. It is assumed that the film is firstrate. That fact alone, however, may not make the sale. Placing a reliable, easy-to-operate projector in your client's office might help clinch the deal. Prepare a presentation kit for your client containing brochures on various types of self-contained projectors. Some of these desktop units offer traditional large-screen projection, while others feature rear projection on small pop-up screens.

Perhaps the most impressive projectors are those using continuous-loop cartridges. They are quiet and extremely easy to operate: one need only open the case and push the start button. Be sure to emphasize the ease of operation to your client: it is a strong selling point for both the projector and the filmed demonstration concept. If your client buys a continuous-loop projector, have your film loaded into cartridges by a local audiovisual lab or one of the national cartridge-loading services.

I do not recommend using a continuous-loop projector for showing your own sample film for two reasons. First, open-reel projection is faster and more convenient to handle should the film break during your presentation. Second, it will be simpler for you to make any necessary deletions or additions to the film.

If using video, be prepared to display all formats, including 8mm videocassette.

Making your client's film is only half the job. Insuring that it can be properly displayed under the most favorable conditions is the other half. If he buys his projector directly from you and you take the sales markup, suggest that he obtain a service contract for it. In many cases he can take advantage of expedited factory service offered by the manufacturer.

Other Potential Markets

Whether or not you make that first sale, you now have a sample reel safely in the can and have survived your first sales presentation. What other businesses can you approach with the sample? Start with the Yellow Pages; you will find all types of potential business contacts for one-shot films. Contact the local chamber of commerce: they usually have lists of all member companies in the area. Although these lists sometimes cost money, they are highly useful.

Do not rule out larger markets. Auto dealers often display action films provided by the manufacturer's marketing division. Local dealers, however, may also sell specialty vehicles or trucks that have applications in the field—applications best demonstrated on film. Perhaps they want to show off their clean, reliable service departments. A film of the service crew at work will eliminate repeated disruption of their work, as well as the possibility of customer injuries and related lawsuits.

Use your imagination when looking for prospects. Cabinetmakers may want filmed samples of their work—major jobs that were done for out-of-town customers. Plumbers may want films of complex and unusual installations. Bakers of specialty wedding cakes may wish to demonstrate their craft.

Cost is the most common objection to using film. Try to start with low-cost offerings. Even if the budget is very small, you have the equipment and time, so use them to record different images for a variety of clients. Introduce your prospects to the use of film or video. When they realize its utility, impact, and immediacy, they will know it is worth investing in on a scheduled basis. One contact will lead to another. You may not get rich, but a good living is there waiting for you. Just reach out and grab it.

4

SELLING

We sat in a French restaurant a couple of blocks from the Film Center Building long after the lunch crowd had gone, long after the dinner tablecloths had been laid down. My longtime associate, Ernie—once a power in Manhattan filmmaking who was clearly headed for Hollywood with a great high-adventure property under his arm—was now folding.

"I'm through," he said for the fourteenth time, as if trying to convince himself.

"What about that book you were going to write? Remember, the one about editing called Making Ends Meet?*" He shifted uneasily in his chair.*

When we finally parted, I watched Ernie disappear into the dark crowd on the street. It was almost as if he had been swallowed up by the horde, which was so thick it had a life of its own. At the same moment, a young woman carrying a film can walked by briskly. Lively and ambitious, she snapped her fingers as she walked. What a contrast to Ernie. He was defeated: she was not yet acquainted with the word.

To succeed in filmmaking you must never succumb to defeat. Whether you are just starting out or are a veteran of many years, you must maintain a fresh and eager attitude in your approach to the work. Headhunters throughout midtown Manhattan describe those with this ability as self-starters. If you are a self-starter, you are never truly unemployed. When you are not working for someone else, you are simply working for yourself. Your outlook remains positive.

Successful people do not look toward the outside world for incentive: they turn themselves on. It is simple enough. They have learned how to rev their own engines before hitting the track. If you have or can develop this ability by using a series of positive thoughts, you will succeed in selling your talents.

THE BASICS OF SELLING

Understand from the very start that *sales* is not a dirty word. Good salespeople are honest, well organized, and masters of their trade. They are also experts in their prospective customers' businesses. Astute salespeople plan well and know how much "sell" to apply. They know the difference between a hard and soft sell and who gets which. Successful salespeople keep good records and understand the meaning of follow-through. They are sincerely interested in helping their customers and do not abuse their talents for persuasion. The work is hard, tedious, and—for those who cannot separate ego from business—often humiliating. Good salespeople are, however, among the highest paid professionals in business.

Nothing moves in business without first being sold by someone. If you have ever taken a telephone order, do not assume there was no sell involved. The sell took place before you received the call. Perhaps your customer sold himself, but the sell was there.

As an independent filmmaker or audiovisual producer, you will be selling an idea rather than a product that can be seen and touched. Only after the idea is sold does the tangible product—the film—become a reality.

Techniques for selling the idea of film are presented in the following pages, along with a brief description of the three basic phases of selling: the *canvass*, or solicitation; the *negotiation*, or actual sell; and the *close*.

The Canvass

Canvassing is as tough as you make it or as much fun as you want it to be. If you assume at the outset that you are an intruder on your prospect, you will certainly fail at canvassing. Your prospect will sense your lack of confidence. Instead of perceiving you as an exuberant flash of dynamite, your potential client will consider you a waste of time. The sale will be lost before you utter a word. On the other hand, if you are convinced you can help your prospect sell a product or service because you are a good filmmaker, count yourself in as a canvasser. The image you project implies that you have done your homework and know how to ply your trade as a filmmaker efficiently and honestly. You can be trusted to use appropriate means to promote your client's business and produce a film for the smallest possible budget. That is really all there is to it.

The key to successful canvassing is confidence. You become confident by knowing not only your own business, but also your prospect's business and communications problems as thoroughly as you can. It is a simple process: All it requires is a willing attitude and a positive mind set.

How do you canvass? The success of your approach to a prospective client will depend on your preparation for the encounter. You do not call your prospect and ask, "Would you like a film?" Instead, consider the following dialogue as one effective approach to an unknown prospect:

FILMMAKER: (Dials phone number, takes deep breath, and smiles.)

PROSPECT: (gruffly) XYZ Corporation.

FILMMAKER: Good afternoon. This is Toonerville Productions. May I speak to Mr. Smith? (Your research has uncovered Smith as the sales promotion manager, public relations director, or marketing manager.)

PROSPECT: Smith here.

FILMMAKER: Good afternoon, Mr. Smith. My name is Paul Jones. I am the vice president of Toonerville Productions, and I would like to have a chance to meet you and tell you about our business. Can we have lunch this week?

Analyze this approach. Before he makes the call, the Toonerville representative assumes a positive attitude. Using an acting technique, he psychs himself positively by thinking of something that turns him on. He ignores the gruff reply he receives from his prospect and continues to stand his ground, remaining cheerful and courteous. He acts, he does not react. Realizing that people can "hear" a smile on the phone and that emotions are extremely infectious, Jones keeps smiling and tries to infect his prospect with positive vibrations. He is soft-spoken but sure of himself.

Jones properly identifies himself and his company; he also gives his title. In many communication houses and advertising agencies, the title of vice president is used extensively, not so much for ego-boosting as to elicit professional courtesy from the client. A title indicates the caller's career stage and the importance of the business he conducts.

This first conversation is an icebreaker. The filmmaker does not attempt to sell anything; he has simply opened a dialogue between himself and his prospect. He, therefore, avoids giving the prospect a chance to say no. The only question he asks is in the form of a lunch invitation. An invitation to a stranger for lunch is best phrased in terms of sharing, although the film representative plans to pick up the tab.

You will note the filmmaker's invitation pinned down a definite time. He did not ask, "Can we have lunch sometime?" He knew the prospect could answer, "Yeah, maybe sometime. . . ." thinking sometime in the year 2800. Should the prospect respond, "Can't make it this week," the filmmaker is prepared with an alternative: "How about next Tuesday or Wednesday?" Again our filmmaker is savvy. He knows the advantages of offering two choices rather than one. When one choice is offered, the prospect can say yes or no; with two options, the answer no is farther away. He also avoids suggesting Mondays, which are generally rough business days. Typically, the middle of the week is less encumbered, but take your prospect's business into consideration. If you pursue your invitation carefully and suggest a convenient time, the prospect will usually accept the date to avoid being discourteous.

When you ask for an appointment, do not accept a put-off. Be prepared to respond to the objection "We're not interested." Consider the following rejoinders:

"You'll want to see my work in the event an associate asks you about local filmmakers at some future time."

"Let's have lunch anyway. You may be interested at some time in the future."

"I'd like to know more about your division and its marketing training efforts: maybe there is some way I can help."

"I'm very interested in showing you how ABC Corporation solved a problem using film after I offered them a choice of solutions."

Each of these responses meets two criteria: it is positive and it offers the promise of benefits. By answering objections firmly and positively while expressing an earnest desire to provide a professional service, you should be reasonably successful in arranging an initial meeting.

Although few sales may result from these first encounters, after the lunch, the filmmaker and production company are no longer unknown to the prospect. A basic informational exchange is the purpose of the meeting; but if the filmmaker has done his homework regarding his prospect's business, he can usually suggest a filmic solution to one of the prospect's business problems during the course of conversation. Subsequent meetings and negotiations for a film can follow if the prospect is intrigued. If not, you have made your presence known and perhaps have planted a seed that will grow and bear fruit in another season.

Negotiation

After your initial meeting, you are in a better position to assess how film can help your prospect's business and to use this knowledge to sell a film. Does your prospect sell a piece of machinery that can be displayed on film? Is the company manufacturing a new product that needs to be introduced to its dealers? Has an old piece of machinery been given a new lease on life with a new type of application? Does your prospect provide a little-known service that needs more public exposure? Is your prospect aware of how little it costs to present a public service film on local television? Such information is the ammunition used in negotiating a film production.

Present yourself as a professional. A good filmmaker will assure his client that he is providing a service. All appointments should be kept punctually. If a delay is unavoidable, the client should be informed in advance. Paying attention to detail gives the impression that you are organized and will handle your client's affairs meticulously, as indeed you should.

If you are successful in selling the idea of a film to your prospect, the next step in negotiation is usually a presentation. The filmmaker will screen a good sample reel and will be prepared to answer questions concerning the budget for each production represented on the reel. A vague recollection of a budget can destroy a prospect's confidence in the filmmaker.

Following the sample-reel screening, the filmmaker will offer a formal written presentation if the client has agreed to accept one. This presentation, referred to as a proposal, usually contains the following: a statement of the problem, a rationale for the solution of the problem, the cinematic solution to the problem, and a comprehensive scenario of the forthcoming script, plans for distribution of the film, a comprehensive budget (covering as far as the first answer print) with a grand total figure.

The proposal is discussed in detail later in this chapter. For now, it is sufficient to note that a proposal should be flawlessly typewritten and put in a presentation folder (available at most stationery shops). For a professional touch, the firm's name, address, and telephone number—with its logo, if any—can be imprinted on the folder, or a label printed with the company name can be affixed.

If you advance a bid at this point, it must be very comprehensive, covering every step of the production. You must not only justify the bottom line against the client's expenditure, but you must also balance it against projected sales or growth of business. The client must know how the money will be spent.

If your proposed film is a public service film with a news angle, do not be timid about proposing that film clips or a story be offered to a local television news show, a local newspaper, or a trade magazine. Your clients' public relations departments will do this type of job, but it helps to be sure they are aware of the possibilities. If your client has no public relations staff, include public relations in your proposal. Also assure your client that a good distribution plan will be created for the film. If you do not want to be directly involved with distribution, acquaint your client with professionals in the film industry who can help.

In every step of the negotiation, the filmmaker's preparation will prove the key to success. Those able to combine knowledge of the filmmaking art with an understanding of the functions of advertising, sales promotion, and public relations in business, will undoubtedly find themselves closing a film production deal.

The Close

If your client accepts your proposal, you are ready to close negotiations. Never close with a handshake alone; even if you have a long-standing relationship with the client, you should formalize your agreement in writing. All business transactions should be finalized in written form. If the production is small, a comprehensive letter of agreement signed by both parties is sufficient. In all other cases, you will need a contract.

The contract is generally provided by the film production company. As discussed in Chapter 3, contract forms are available at film supply houses, but it is best to draft your own contract to suit your needs. The contract should detail precisely the terms of your agreement with the client, outlining not only the filmmaker's responsibilities but also those of the client. An attorney should review your contract design. In some cases, you will want an attorney to draft a separate contract for a specific client.

Follow-through

Once the contract is signed, your sales work is completed. Or is it? You are now dealing with a potential repeat customer. Even if your clients never contract for another film, they have contacts in the business community. Their impression of you and your work will be what is passed on to their associates and colleagues. Ask yourself, "What can I do to assure that my clients will sell for me?"

In following through on a successful sale, the first step is obvious—delivery of a first-rate product. After that, assure a continued relationship by calling the client occasionally. Inquire how the film is doing out in the field and what, if anything, you might do to assist him in the distribution process. It is not necessary to solicit repeat business directly. Remind your clients of your sincere interest in their businesses periodically, keeping abreast of changes and developments. You will then be in an excellent position to negotiate when a new opportunity arises for another film or video solution to a communications problem.

PREPARING A FILM PROPOSAL

Just as follow-through can assure repeat and new sales for the filmmaker, a well-conceived proposal can clinch a deal, providing your client with a clear picture of the finished product as well as justification of the expenses involved. Bearing in mind this important fact, take a closer look at the basic parts of the proposal along with an actual rationale for a film with a $90,000 budget.

Be aware that what follows can serve only as a model. In most cases where a proposal is requested, detail is encouraged. Indeed, some clients weigh proposals by the pound. Such detail is most often developed in meetings with your client designed specifically to elicit the information required to add substance to your proposal.

Outlining a Film Proposal

Prior to drafting the proposal, take time to research your client's business. Evaluate the client's situation during initial meetings. Does the client have a problem? What is it? Does the client understand the problem and do you both agree on what it is? What are your objectives? Before you put a word on paper, you must feel confident that you not only comprehend the client's problem but also have a viable solution to it in the form of a motion picture. Only then should you proceed to draft a proposal incorporating the following four parts:

1. *Rationale*

   Define your problem. Although your client may have initiated the project, others might make the final decisions. Help your client and yourself by writing a concise rationale stating the fundamental reason for the film. What will the film do? Will it change the audience's behavior? Will it result in a change in the client's market penetration? How? Will it introduce a new service or product? How? What are the projected results of the film's use? In barest outline, a rationale should cover the following points: presentation of the client and definition of his or her problem; identification of the intended audience; the proposed solution to the problem; an explanation of why film is the answer.

2. *Treatment*

   A treatment is a very tight outline of the script and film (or in audiovisual productions, the visualized message). Determine how much can be spent on each aspect of the film before writing the treatment. If you are working with a small budget, forget expensive location jaunts. The treatment of a low-budget film calls for crafty as opposed to lofty imagination. Include the proposed running time. Remember that films are for impact and should not be burdened with details that befuddle the viewer. Length should also be determined by balancing the film's concept against its budget.

3. *Distribution*

   Suggestions for the film's distribution should be included in the proposal. What is the expected method of distribution? Will the film be transferred to videotape or Super 8 for salespeople's use in continuous-loop cartridge projectors? Will it be shot for 35mm blowup for theatrical distribution? Or will it be shot in 35mm and reduced to 16mm or transferred to tape? When you design and build a vehicle, which is exactly what a film is, you should know the distance and to what destination the vehicle will travel. Budget considerations also play a role here. The filmmaker must be able to predict not only the cost of distribution, but what it will cost to transfer a print to another format, film gauge, or electronic medium. If the film is to be used effectively, the client must know at proposal time how to use it.

4. *Budget or Cost Estimate*

   The final part of the proposal answers the question most important to your client: How much will it cost? Prepare a list of expenses based on your treatment. Determine the degree of detail your client requires in an estimate, and prepare your budget with a built-in contingency for strikes, inflation, and natural disasters. If you need help, refer to the sample budget in Chapter 3. When you have completed your draft proposal, do not be shy about discussing the proposal at length with your client. Where necessary, make changes according to the client's needs. After all, only client approval of the proposal will culminate in a signed contract.

The Rationale: A Professional Example

Although the client's first interest may be the budget, the rationale is the real heart of the proposal. If the rationale argues its case effectively, the expenditure for a film can be justified in the client's mind before he reads the bottom line of the budget.

The following example of a motion picture rationale was written by John Hammond for a large northeastern farm cooperative here called Grange Farm Enterprise Service. The film was produced to be shown by the client directly to his audience and for airing on local television in a strategic Sunday afternoon slot.

PREMISES

The primary audience of the projected Grange Farm Enterprise Service motion picture has been defined as 30,000 large farmers in Grange territory, of whom 10,000 are currently substantial users of Grange products and services. In designing a motion picture for this 30,000-member "prime" audience, it is fair to assume the following premises concerning these individuals: Anyone who has built and is successfully sustaining a large farm enterprise today is almost certainly an intelligent, competent individual who very possibly could be realizing greater income from some other kind of enterprise if income were this individual's only criterion of business and personal success. The attractions of farming that have the strongest appeal include the relative independence and self-reliance afforded by the farm enterprise.

SUGGESTED THEME

If these premises concerning the major target audience can be accepted as valid, then an effective theme for this film would be "Independence through Cooperation." A more specific contention would be that Grange Farm Enterprise Service is the first full manifestation of the true cooperative principle. The goal of the cooperative has always been the greater success of the farm enterprise, but up to now it has been manifested only in partial aids and solutions, such as the low-cost or quality-assured commodities.

Enterprise is the culmination of the cooperative—a dedication to the overall success of the total farm enterprise rather than merely seeking to improve isolated parts of it. Correlation and synthesis of the vast number of inputs involved in this total enterprise approach have been made possible by aggressive utilization of the latest techniques and tools of information technology and management science. In the face of the proliferation and growing complexity of farm-information inputs, it is resorting to the help of information technology and management science that will enable today's farmer to remain independent—to continue to make decisions based on better and more refined information and to sustain independence through a pooling of resources with other farm entrepreneurs "to do together the things we individually cannot do (alone)."

SOME FACTS OF LIFE

To be effective and credible, this film must reach its audience in a context of realistic recognition of the way things have been up to now and their anticipated status at the time of the film's showing. Following are some of the facts that the film should recognize and address.

1. In attaining the size and strength inherent in having the resources to meet the needs of today's farmer, all agricultural cooperatives (not just Grange) have encountered a problem of alienation from their membership. Many farmers feel that size had led to a loss of consciousness of the basic purpose on the part of their cooperatives and that the co-ops have forgotten who is working for whom.
2. Today's farmer is caught in a withering cross fire of reports and recommendations arising from the work of some 10,000 agricultural scientists and researchers. Some of this information is (or seems) contradictory; all of it needs correlation to the farmer's individual situation.
3. Farmers—in common with most individuals—generally misunderstand the role of the computer in management decision making. They may resent a computer telling them what to do.
4. The Enterprise Service representative is a normal individual who possesses no extraordinary personal powers beyond good native intelligence and a conscientious resolve to make the maximum possible contribution to the success of his Enterprise Service clients. No one should make the mistake of inferring that the Enterprise concept relies for its efficiency on any fundamental internal change in the Enterprise representatives themselves.

COMMUNICATIONS OBJECTIVES OF THIS FILM

1. To establish that the Enterprise concept is a major new phase and really the culmination of the cooperative movement—this time seeking to foster maximum success of the total farm enterprise rather than simply seeking to improve the economics of isolated sectors of it. This is what farmers were really seeking to do when agricultural cooperatives were started in the first place, but of necessity could deal only with the individual facets of their problems. The new "total" Enterprise approach has been made possible by breakthroughs in information technology and management science. (The suggested title, *The New Field,* alludes to the fact that Enterprise seeks to be a rededication to the original purpose of the agricultural cooperative.) By the utilization of those management aids—made possible by the pooled strength of the cooperative—the progressive farmer will be able to retain individuality and decision-making independence in the face of

the growing complexity and pressure for increased size that characterize farming today.

2. To make clear that one of the goals of Enterprise is to make the Enterprise representative more effective in the farmer-user's behalf by making available extensive information and information-analysis resources, as well as by increasing knowledge of the existence and use of these resources. The value of the representative to the farmer is as an information and consulting "terminal" plugged in more firmly than ever to vast data and analysis resources at Grange Headquarters.
3. To establish that the success of Enterprise Service depends heavily on the placing of complete confidence by the farmer-user in the Grange representatives with the same assurance with which personal financial information is disclosed to an attorney, banker, or real estate counselor.
4. To convey that the role of information technology in general (and computers in particular) is not to "tell managers what to do," but to provide more highly refined information and staged series of management alternatives as aids to more intelligent decision making by the farmer-manager.
5. To communicate unequivocally that Grange—as represented by Enterprise representatives through backup specialists to top management—regards itself as a new, fuller manifestation of the ultimate cooperative purpose: contributing to the optimum success of its member enterprises. This film should make clarion-clear that Grange, at least, is one cooperative that has not lost sight of its reason for being and that Enterprise Service represents an opportunity for a new and fuller partnership between Grange and its members.

WHY A MOTION PICTURE?

The effectiveness of this communication is going to depend as much on the portrayal of attitudes, apprehensions, empathy, dedication, and the need for trust and other emotional elements as on the conveyance of a good deal of factual material. It is difficult to conceive of any other medium in which the nuances of image, motion, expression, voice inflection, and visual and audio effects can be deployed as effectively—and as unalterably—as in a motion picture. Once complete, a motion picture will be a more "official" statement of the Grange's aims for Enterprise than would any other visual medium.

A motion picture encapsulates power, impact, life itself. It freezes a moment of time, retaining all the vigor and drama that was professionally designed into it to be retrieved at any moment desired.

If you have merely glanced over the preceding rationale, go back and study it seriously. Although a rational does not have to be this comprehensive, it must reflect solid research and understanding on the part of the filmmaker. The rationale indicates the filmmaker's awareness of the client's problem and defines a solution. It also sells that solution by using creativity and fresh ideas.

It should be obvious from the Grange example that any proposal must be written in clear, precise prose. If you are unable to communicate your ideas on paper, hire a professional writer to translate your understanding of the job into a proposal your client will accept.

Above all, do not underestimate the power of a well-conceived and professionally presented proposal. It is the nucleus of your sell. Even if your canvass yields many prospects, only an effective presentation/proposal will guarantee a signed contract and the opportunity to follow up that initial film with repeat business.

5

CLIENTS AND CLIENT STRATEGY

Joe Smith is a good filmmaker. He works fast and well, coming in on deadline and under budget every time. Joe looks good, watching his diet carefully, selecting fresh vegetables, fruits, and grains with the precision of a Toyota quality-control expert. Yet he does not govern the quality of thought he allows to flood his consciousness. Just when you think he might be on top of the world, Joe is depressed and withdrawn. He pursues negativism until it drowns him with anxiety and remorse. Joe feeds his mind with garbage.

Part of Joe's negative syndrome is his inability to think himself out of a small restraining circle, to reach beyond the abstract wild blue into the wild purple. The very answer to Joe's dilemma is often just around the corner, but he pauses and abandons his search out of sheer frustration. Always remember that your oasis may be just beyond the next hill, your next sale might come with your next phone call.

As a filmmaker reading this book, you and Joe share a common goal: to make money in film. Your success in this venture depends largely on your ability to attract clients. Fortunately, the range of clients available to filmmakers is diverse and constantly growing, but this very fact increases the variety of sales tactics required to capture money-making contracts. You must know in what direction you are headed and keep your wits about you.

The selling basics described in Chapter 4 will provide the foundation for most sales situations you encounter. What follows here builds on that foundation by offering a description of three different roads to making money in film and of a strategy for building your business in each area.

PUBLIC RELATIONS

If you are looking for a film career that offers a fair living without the complications of backbiting egos, complex production techniques, and excessive competition, you may want to investigate film public relations. Although traditional public relations relied principally on press releases and feature articles in the print media to inform the public of corporate developments or institutional problems and changes, recent awareness of the impact of the visual media has changed this emphasis and opened an area of opportunity for the independent filmmaker.

Although network crews and local stringers amply supply hard news, many local television stations are still delighted to receive short segments of film portraying corporate, institutional, or organizational news. Such clips are used to fill time requirements for local news and even otherwise unprogrammed time.

Public relations clips must have a news angle. A new product that expedites a manufacturing process is not newsworthy: it lacks direct public impact. If, however, that product directly benefits the public—by enabling the phone company to make repairs from headquarters and, thus, eliminate 85 percent of their service calls, for example—it is news and a proper subject for film.

Soliciting Work

To get started in this field, you will want to canvass both your potential clients and potential outlets. The business desk at your local library can help you compile a list of all the public relations agencies, corporate public relations departments, news bureaus, and television stations in your area. Begin with the public relations professionals. Spend several weeks meeting them, soliciting their advice, and exploring the possible opportunities for using film to deliver their messages. Then proceed to the local stations. Find out about each station's policies on filmed press clips/releases. Most will be willing to consider such offerings. If a station is not interested, put it on a list to be contacted again after you have a track record. Its attitude may very well change.

When you have completed your canvass, ask for an assignment from one of the public relations pros who might have encouraged your endeavor. Your first job will probably come from a public relations agency, unless the director of public relations or the public information officer of a large corporation in your area gives you the go-ahead.

Once you have landed a contract, it is your job to keep your client's interests in mind. This may be difficult, since public relations people are driven by everything from rampant self-interest and flat-out misrepresentation to solid devotion to trivia. You may be asked to film a spot about a product that insiders feel is newsworthy but you do not. In this case, you will want to check with your media contacts to determine whether your clip will get air time and then negotiate. Savvy, tact, and your negotiating skills will inevitably come into play. You can avoid becoming a hired flack by exploring far-reaching areas and cultivating every public relations source that may hand you a story.

Production Tips

Before you begin your first mini-opus, be sure you understand each station's guidelines regarding clip lengths (usually short, but get specifics) and type of sound track, if required. For material submitted to a local news service, it is quite likely that the news director will prefer no voice-over. In such cases, the film will be delivered with a written release bearing all pertinent information and the telephone number of the releasing party to be contacted for verification or further information. Knowing these broadcast requirements in advance can save you production time and money.

On any assignment, your job is to film the news piece in the most interesting manner possible. This is typically accomplished by portraying the actual situation using a single system and cinema verité techniques. You will want to use all the film tricks in your bag;

a novel change of angles may add spark to an otherwise mundane subject. Remember, the creativity must come from you. (Budgets will rarely allow renting copter-cum-mounts for aerials, cranes, or other fancy paraphernalia.) On the other hand, do not overdo it. You are shooting for the nightly news, not Hollywood.

Once again, an old fashioned single-system camera with a magnetic stripe on the film stock, if you can find one, is fine for public relations spots that are going to be projected onto large screens. For television and videocassettes, obviously, any kind of video format is fine as long as the quality is professional. If possible, do not use amateur equipment. The resolution simply is not there, particularly for reproduction work.

A double system (camera and sound equipment separated) means much higher cost for public relations work and is used only when making a TV special or public relations film with a healthy budget. Most of the PR work you can get to keep you going is in the blurb area, for example, newspegs and new-product or personnel video for TV news usage. For this type of work, the video gear is always least expensive, and, with the immediacy, you will need to get your material into the evening news within the same day for the next day's news.

The best approach to becoming a "supplier" or "stringer" who also works the PR beat is to talk to as many TV newspeople as possible and get them to know you. Do not stop with the anchors and other on-camera people but spend most of your time around the newsroom with production and editorial people. They are the ones who will finally provide your passage into the station news team as a stringer.

Although this section began by pointing out the money-making possibilities in public relations film, once again, you should not plan on getting rich in public relations film and video. You cannot charge an advertising-spot rate for news clips because public relations budgets are smaller than advertising budgets. Business generally believes that public relations is less effective than advertising and, thus, feels it should pay less—and does. You should still negotiate a fair rate for your work and should, by all means, be sure to cover your costs.

The public relations professional for whom you are working will most likely place the film into the media. Should you do the placement, you are entitled to a fee in addition to your shooting fee. Negotiate this. You should get, at least, a top hourly rate for time spent placing it. Several times your normal hourly rate is quite fair. Your compensation should be increased accordingly if you can successfully place the story with a national bureau or network, broadcast or cable, such as CNN, which uses and always looks for much material that is newsworthy.

If you are drawn to this field, study as much as possible, and join the various public relations groups in your area. The largest is the Public Relations Society of America.

One additional money-making avenue is open to the public relations filmmaker: a major film story can emerge from a filmed news release. Quite often a release may hold the grains of a story that can be developed into a ten- or fifteen-minute promotional film for the company. This will not only benefit the company, but it is also an added source of income and exposure for you. So keep your eyes open and pursue such opportunities.

Nurture your business by maintaining an extensive contact list. Befriend the media people (without ingratiating yourself) because they are your bread and butter. The XYZ Corporation may be paying for your film, but it is the LMN media that give you the break by airing it.

Business will also prosper as your technical abilities increase. Study newswriting techniques and read everything you can about the public relations game. A good book on reporting will broaden your perspectives, although reporting is not your primary objective. The books on these subjects mentioned in the Bibliography can be found at most libraries and are well worth your time. Unfortunately, very little has yet been written on film/video journalism.

The field of public relations film has yet to be fully exploited. The idea must still be sold not only to businesses and organizations with newsworthy information, but also to some segments of the broadcast media. The time has come for public relations film, and the filmmaker who can present the opportunities of this field properly—with patience and professionalism—can succeed.

YOUR COMMUNITY AS A CLIENT RESOURCE

Beyond public relations, every community has a second area of prospective clients often overlooked by independent filmmakers: local institutions, government agencies, and community organizations. Hundreds of these entities exist and need to communicate with the public. However, they need to be informed about how film can help them, and they need to be contacted before you can sell your services.

Educational Films

There are still films being made for educational markets, but this industry is a tiny fragment of what it once was. Today, most of the so-called "educational films" are made on video. These will be covered in Chapter 9, which is completely devoted to video.

Religious Markets

Religious filmmaking has never been a very big field, but popular religious videos have made a lot of money, most notably the various versions of animated Bible stories. There are also live-action Bible stories and always have been, but the best of what is properly called *religious films* are made by the communications offices of the larger churches. Most notable of these have been from the Church of Jesus Christ of Latter-day Saints, in Salt Lake City. Obviously, unless you are recognized by the various churches, your chances of making films for them are nil.

If you are an active member of a large church, check with the public relations or communications office on a regional or national level. However, note that each large church has its own production facility that employs gifted members of its congregation. Other fine religious films are produced by the news departments of the television networks, but the shops are closed and offer few prospects for the independent.

For those who feel they really want to do this type of work, consider working for a film producer established in the religious-film field to gather knowledge about the industry, or join a denominational production company. Once you have acquired credentials and developed contacts and a firm understanding of the field, you may be able to hang out your shingle. Consider also the subcategory of spiritual films, which are essentially nonsectarian and often secular in content. The old "Insight" television program, a half-hour show that illustrated reason and morality in everyday life, was this type of production. A continuing market can be found for inspirational films, and if yours is well written and conceived, it could stimulate perennial demand.

The Community at Large

Although the money-making potential of the educational and religious markets is limited, less obvious markets in your community can be cultivated to generate a healthy income from film. Develop contacts at the chamber of commerce, the local advertising club, and the trade organizations. Present yourself initially as an interested individual rather than a filmmaker. Consider the crime reporter who gets his best stories by hanging around the police station and the firehouse striking up conversations. You can use the same tactics to determine where potential film business may be found in your community.

In some cases, a film will be financed by individual rather than organizational funds. A film that I produced called *Rochester, First Person Plural* was made for the incoming president of that city's chamber of commerce. The film stated his philosophy and what he hoped to accomplish during his term, while enlisting support for his projects. It became the keystone of a new program that served the community well.

Aside from business organizations, attempt to establish contacts within city, county, state, and federal government offices. Consider the broad range of possibilities: the school board, armed forces recruitment center, sheriff's office, and department of social services. A complete list is readily available in the local government reference pages in your phone book. Also look into quasi-governmental agencies such as child-abuse centers and libraries. Do not forget action groups like environmental protection organizations in the private sector. In each of these areas, the need for films exists.

When you approach an organization, ask them about community relations. Some will have a promotional budget and others, such as the library, might have a patrons or "friends" group with a budget that can be used for filmed communication. Many of these groups have never contemplated using film, so your interaction with them is crucial. Make yourself known to them and stress that your intention is to help by using the skills of your profession. When a communication need arises, your name will be remembered. Establish the network of contacts first; the opportunities to sell will naturally follow.

If you are interested in government jobs, be prepared to bid on each project. Your first step will be to request a bid application from the contracting agency. Then, using all your technical and business skills, you will prepare and submit a bare-bones budget. In this game, it is essential that you have a positive attitude. Recently a filmmaker who works almost exclusively on government accounts admitted to me that he encountered little competition in bidding. That did not mean he did not carefully watch the bottom line; rather, he felt that few filmmakers bothered to submit bids because they assumed they did not have a chance. To say it one more time, you cannot make any money unless you get out there and try.

LARGE-ACCOUNT STRATEGY

Despite the importance of contacts in attracting new business, most filmmakers err by relying solely on contacts or referrals to keep their businesses rolling. This course is inefficient and risky, yet it is pursued because it is easy. Successful filmmakers advance beyond this simple method to develop the obvious next step: a sales strategy. Indeed, if your sales target is a large corporation, a contact alone will not assure you a contract. Only a carefully planned effort will allow you to determine and then convince your client of the important role film can play in attaining corporate goals. Chapter 11 will cover more on corporations.

Research

Before approaching any corporate contact, you should find out everything you can about that corporation. Quite often all the material you need is at your local library. After reading a corporate annual report, you should have a firm grasp of the company's product or service as well as an awareness of possible areas for corporate diversification and expansion. These facts are of particular interest to you since they will be the subject of corporate communications, and some of that communication will be on film. If you do not have a contact, the report can also provide the names of potential contacts: the training director, advertising director, and public relations manager. You may even be able to glean some information about your potential contact's educational and professional background.

While doing your research, bear in mind that the large corporations composed of multiple divisions producing diversified products—for example, General Motors or Lever Brothers—will have decentralized buying and budgets. Study each part of such conglomerates; film possibilities abound. A need may exist for a corporate orientation film for all employees, plus several additional training films specific to each division's manufacturing or service problems. These films, however, will be purchased by different executives, so you will want to find out who they are.

All research should be thorough and methodical. Commit yourself to legwork and time at the library. Quite often, a letter to a technical, trade, or professional organization can provide in-depth information not readily available at the library. In addition, do not hesitate to contact the company's public relations office for information. Such departments are pleased to provide information to prospective suppliers.

Preparation

Prior to approaching a corporate prospect, check your work samples and reels for length, quality, and appropriateness. Your research will be useful in choosing the right samples. As you determine your client's needs, you will want to specialize your reel for that particular function, product, or service. A training director will be unable to appreciate your genius as a 30-second television-spot wizard, nor will an advertising agency producer be impressed by your 15-minute film on breaking down a thermonuclear reactor. Using your knowledge of your potential client should allow you to prepare your reel and pitch to suit that individual's needs.

Approach

The course of the sales meeting will generally follow the scenario conveyed in Chapter 4. You must always approach your prospective clients in a positive and professional manner. Using the following guidelines will undoubtedly enhance all encounters you have with your clients.

- Never underestimate your prospect's intelligence or understanding of what you do.
- Never press on relentlessly. If a prospect seems harried or busy, conclude your phone call but try to contrive an opportunity for a second call in the near future. Know when to push and when not to push by being a good listener and observer. Watch for signs of fatigue in a meeting. Clock-watching is one surefire danger signal. A prospect has work to do and may not have any more time for you.

- Never deceive your client or prospect. If you have developed this style, break the habit before it breaks you. Deception, charming or otherwise, is poor cement in a relationship.
- Never project a negative stance—whether it be toward the prospect's competition, an individual, or the state of the economy. Optimism is always best. Remember, also, the power of a smile.
- Finally, avoid doing all the talking. Learn to be inquisitive and a good listener. It requires effort, but the information you pick up by listening could very well help you clinch a deal.

Follow-up

After your initial meeting, you should be able to determine the potential for film business within the corporation. In making your projections, consider the following factors: corporate growth during the past five years; net profits for the past five years; advertising and other communications expenditures for the past year; diversification of operations, products, or services.

Another variable is the relative sophistication of management in the use of new communications technology and the recognition of its importance. The filmmaker can help control this variable by educating management before selling. Quite often the manufacturers of videotape or a film distributor will share the expense and effort involved in this educational process. In other cases, your point might best be made by screening a successful sales-promotion film produced for your client's competitor. The most difficult part of your job can be convincing your client of the impact of film.

If you determine that potential business exists for you in the company, draw up an organizational chart to monitor your various sales approaches. Include all corporate officers from the marketing director on down, drawing distinct lines up to supervisors and down to subordinates. Pencil in public relations personnel and anyone who does training. Under each name, place pertinent information such as phone number, extensions, and notes on personality. You will also want to keep a log of all phone calls to the company and note the results. Such steps can allow you to coordinate your sales approach to attract multiple contracts from a single corporation.

Throughout this chapter, potential markets for the filmmaker's work have been surveyed, but most of the discussion has centered on business practices adaptable to almost any profession. Why? Because to make money in film, you must be a good businessman.

Filmmakers begin with cinematic skill and talent. Continued success depends on detailed planning and careful handling of accounts. By marrying your art with sound business practice, you can make good films and a good living.

6
SURVIVAL

Although this edition is dedicated to Jamison Handy, whom I cover in Chapter 12, the first edition was dedicated to the great director Frank Capra who died during the preparation of this edition. I felt it important to keep the background on this extraordinary director and human being as the introduction to this chapter, as it was in the first edition.

In April 1981, Esquire *published a letter I wrote to them in response to a positive but superficial article on Frank Capra that they had recently run. In part, the letter read:*

> *Mr. Capra was the first to give a skinny and hungry cartoonist—Walt Disney—a big break when the struggling animator approached Columbia Pictures during his first attempt to sell his own product. Frank Capra had walked into a then-third-rate studio (Columbia Pictures) and, with his brilliant engineer's mind and his human heart, brought it to fame and glory.—As for Frank Capra's not making a film in the last 20 years: that is perhaps one of America's greatest tragedies. But his films will still be*

> *playing in some distant tomorrow long after those who number among the self-indulgent, talentless, and illiterate directors of the present are gone. They will be playing anywhere there are audiences hungry for the beat of the human heart.*

Frank Capra lived his life with the kind of passion illustrated by that letter, a passion that is now, unfortunately, out of fashion for the time being.

When, as a young boy, I first became enamored of Frank Capra's work, I knew very little about him. Actually, I would have had to search hard to locate the few articles that had been written about him. It had been a long time since he and the handful of greats—Ford, Wellman, Wyler, Wilder, Stevens, and the rest—had made craft history not matched since.

Today, you can read all about the great directors almost anywhere. They became great not by wearing designer jeans or by driving exotic cars, but by applying themselves. They made a commitment to dig in and work hard at something they chose to do.

Lost Horizon *inspired me as a boy, and it helped to shape my mind. Above all other films I have ever seen, it gave me a spiritual direction that has lasted to this very day. As a teenager, I sat saucer-eyed in front of the TV watching episodes like "Our Mr. Sun" and "Hemo the Magnificent" from the science series that Frank Capra, with the help of animator Shamus Culhane, produced for Bell Telephone and N. W. Ayer & Son. For the first time, I thought of the great promise of television and its potential for mass education. The films entertained as well as informed, ushering in an entirely new era for TV.*

Throughout my career, I honestly sought out the formula for greatness because, like many other "little" people, I too yearned to be great. As the decades rolled by, I learned to distinguish greatness from fame and began to realize that the essence of greatness was intrinsic to each of us. It had to do with a fundamental concept so well defined by Frank Capra.

In observing Frank Capra's work, one cannot miss his underlying message. He actually gives it away in one of his films, his classic It's a Wonderful Life. *In it, the protagonist, George Bailey's father (played by the wonderful character actor Samuel S. Hinds) has the magic theorem written on a small plaque mounted on the wall behind his desk: "The only thing you can take with you is that which you have given away." If the George Bailey character had discovered this (which was right under his nose) for himself, there would have been no story, no classic film, and the extraordinary James Stewart would have been out of a job.*

In the mid-1970s, I had the opportunity to have dinner with Frank Capra, and I listened to him with the reverence of a school child, noticing that within almost everything he said was a tidbit of advice or information. He never stopped giving. I asked him many questions, but one lingers in my mind because of its significance: "Why did "Gunsmoke" last so long on television? It was just another western, wasn't it?" Without hesitating, Frank said, "Because each week the audience was given a love story. . . . The protagonist loved his homestead, his horse, her little boy, and so forth. People love a good love story best."

Frank Capra knew much about love. His work reflects it from his very early silent films to his features, World War II documentaries, and his early and exceptional television series. His work is mostly humanitarian.

Thus, we come to the key to survival (in this business) and the solution to the mystery of greatness. Through our work, we give not only to others but to ourselves as well. Surviving is easy if you bring love to your work. Greatness is measured by what you can give. Giving is very painless when you realize you are also the recipient.

Having read this far, you should have a good idea of how to get started as an independent filmmaker. You also may have a number of questions about the process. What follows is a series of frequently asked questions with answers derived from my own experience as a commercial filmmaker. They, in turn, are followed by a compendium of tips for surviving in the filmmaking business.

SOLICITING WORK

Q: I work in a small town. Since there are no labs here, would it be impossible to do professional work, even on a small scale?
A: If you say it is impossible, it will be. If you live in a fairly large city—one with a population near one million—that has no labs, cultivate the local television stations. Those that have not switched to electronic news gathering are probably equipped for color processing and small-scale editing. Otherwise, obtain a list of labs in neighboring cities by checking the Yellow Pages of those cities at your library or by writing to the Association of Cinema and Video Laboratories. Select several of these labs and write to them, explaining your needs.

If the response from a lab is encouraging, budget a trip to tour it. (The cost of the trip will be tax deductible if you are doing business in film.) Prepare a list of questions in advance. A good lab is always willing to answer its customers' questions.

Q: How can a lab help other than by processing my film?
A: Aside from film processing and printing, most labs can also provide information on camera equipment, editors, editing rooms,

sound studios, special effects, and completion services. Be certain also to ask who is placing the largest footage orders. The answer could lead to a new client or a "big city" producer in need of your services.

Q: How do I solicit new business?
A: Besides those methods described in Chapters 4 and 5, direct mail is an effective means of stimulating new business. Although it is expensive, some type of 3-D mail is most appropriate for film and audiovisual producers. (Refer to Chapter 7.) Otherwise, it is best to be represented by a professional-looking mailing piece that has been well written—preferably with catchy or clever prose—and strikingly designed. You might wish to hire a copywriter, although you can do the writing yourself, and you should certainly enlist the services of a small art studio for the graphics. If this type of circular is beyond your budget, seek the help of a typing service. A well-phrased letter, neatly typed on your company letterhead and followed up with a courtesy telephone call, will also go a long way toward attracting business.

As stated previously, do not overlook the local television stations. Quite often, they need both news camera operators and budget-priced producers for retailers who buy spot time in your city. They may even be able to provide leads on local businesses—large auto dealers or furniture warehouse outlets, for example—who have requested production services.

Remember that regardless of where you live, finding your market is fun. You will enjoy playing a sleuth, tracking down leads with clues, using your mind, and developing your skill at deductive logic. Do not assume that anything is impossible.

Q: Where do I find clients who are interested in communicating on film?
A: You should know the answer to this one already. Potential clients are everywhere; the only problem is that they are not always aware of their needs. It is your job to sell them on the value of film. The Yellow Pages is a directory of potential customers. Target those firms having the largest ads. Talk to business people about what you know best: film. Tell them about the benefits of film and be prepared to show them how film can save them money.

Remember that you must know your prospect's business before you attempt to provide a film to promote it. Let your prospect know that you are interested, ask about the business, and suggest that if you see a way to help by using film, you will propose it. If you discover there is no business for you, thank your prospect for meeting with you. All is not lost; you have made a favorable contact that might result in an indirect sale.

Q: I have a good reel of sample productions. By that I mean a single reel, and I do not want to risk sending it around. What do I do?
A: Every business, trade, or professional organization has an occasional need for luncheon speakers. Speaking engagements provide an excellent opportunity for you to present your sample in person. Offer to make a presentation on your services to the local advertising club or Rotary. Prepare an interesting talk. By keeping it brief and direct, you can increase your effectiveness and maintain audience interest. If you need an icebreaker, consider borrowing an old-time comic short from the library.

Sometimes you will have to risk sending out your reel, since your work must be seen if it is to stimulate business. Consider making some half- or three-quarter-inch videocassette duplicates for this purpose. Duplicates are quite inexpensive if you have a videocassette recorder; just purchase blank cassettes. If the duplicates must be made for you, the cost is still moderate.

MANAGEMENT DETAILS

Q: What is a simple way of budgeting a film?
A: Review Chapter 3 for a detailed discussion of budget. In brief, there are two steps: 1) compile a list of all production costs, and mark them up not more than the agreed 20 percent for handling; and 2) charge for your time at a rate based on local cost of living. Quite often the individual project will determine your budget. On some, you can choose to forego an hourly rate and work solely for the overhead markup; whereas on more complex shoots with large budgets, you may negotiate a package deal. You will have to judge what is right in each situation.

Q: I just lost a bid to another filmmaker. My bid was as tight as possible. How could he afford to underbid me?
A: Chances are that he could not afford it. You may have been underbid by someone who wanted to capture the account. The low bidder may then try to play "catch up" by increasing costs on any subsequent films he makes for this client. Bidding below your costs is unprofessional, and the end result is a less-than-quality film. On the other hand, if your bid conforms to costs and you come forth with a fine product, you may never have to submit another bid.

If you are really concerned about your competitor's low bid, call him and ask how he could afford to be so generous. If you both gauged your bids on a careful budget and a reasonable profit but his bid was just a few dollars under yours, he probably knows his business a bit better than you do. You will know better the next time.

Q: I understand that most large projects progress from proposal to script to estimate to production. Should I charge for a proposal?
A: If your client requested a proposal, you should indeed charge an hourly rate for preparing and writing it. You may choose to waive this fee when the proposal is the result of your suggesting film as a new technique to your client. If not, obtain your client's approval of the fee prior to billing.

Q: I just received a complex assignment and will need to hire a writer. What is their standard rate of pay?
A: Most writers will quote you their rates. You can also obtain rate information by contacting the Writer's Guild East or West or by consulting the current volume of *Writer's Market* and issues of *Scriptwriter Magazine* and *Writer's Digest*. Writers charge either hourly rates or per page rates. In many cases, the writer will receive close to 10 percent of the total budget for writing a script through to final approved draft. An experienced writer who provides copy and script for advertising often charges $100 and up per approved typed page. Many audiovisual and videotape scriptwriters will work for much less. The trick here is to negotiate. Always carefully verify your prospective writer's references.

Q: How should I invoice the client?
A: As described in Chapter 3, the current practice is to bill film jobs in thirds: one-third upon script approval, one-third upon approval of the interlock, and the final third upon delivery of the first answer print. If your client is a corporation or a large institution, obtain a numbered purchase order detailing the services and products you will provide. Place the purchase-order number at the top of all your correspondence and invoices to identify them and expedite all administrative handling.

Q: Who owns the props, outtakes, and/or material used during film production?
A: It is an industry practice that all aspects of production are owned by the person who paid for them. Unless instructed otherwise, be sure that everything is delivered to your client, including the work print. After the answer print has been approved, send a letter to the lab assigning rights to all property—such as A&B rolls, sound tracks,

and negatives—to your client. Send a copy of this letter to your client; retain a second copy for your own files.

Q: Do I get my name on the credits of my production?
A: Credits for yourself and your staff may be included on the film only with the agreement of your client. It is best to negotiate this point as part of the contract. Avoid giving your client a credit as executive producer. It gives your client creative authority that may not be deserved.

Q: I made a film for a client. He liked it, and it won an award. But I never heard from him again. What, if anything, did I do wrong?
A: Nothing! In fact, that is the problem: You did nothing. Why do you feel that you should hear from him? After all, you are the one who is offering the service. Did you set up a good distribution plan for your client? Did you follow up on its use? Did you check back to see if there was anything you could do to help him or even call just to say hello? If you have not kept in touch, it could be that another filmmaker is harvesting the field you plowed and sowed. You had better pick up the phone right now.

Q: I have trouble collecting my fees. What should I do?
A: First, all work should be preceded by a dated purchase order (P.O.) that is signed by your client. Be sure to get the signature before commencing work. It is for your legal protection and serves as a contract between you and your client. It also protects your client, assuring the work will be paid for and should be completed within a deadline.

P.O. forms can be found in either computer software stores within special form programs or in the larger stationary stores that sell preprinted forms. However, your client should already have them, especially if it is a large corporation or ad agency.

Second, invoicing is always done in thirds to assure that you get at least most of the funds you spent before the job is completed.

Third, if you have trouble collecting, simply keep invoicing every two weeks, sending invoices not only to the signer of the P.O., but also to the accounting department. Locate the name of the person in charge of the accounting department and address it directly to that person, giving all the details, job number, P.O. number, date, description of the job, the name of the person who ordered the job, your name and social security number, your address, and your phone number.

If you still have trouble, make inquiry calls *every* Monday right after 9:30 A.M., or have one of your assistants call. If you have failed to collect within 120 days, then check with the county clerk's office, the Small Claims Court and see if the amount you are billing is within the court's jurisdiction. There is always a no-exceed figure. If your figure is within the amount, file a claim. They will give you all the instructions you need to do this.

Remember that if you begin legal process, you will probably have lost a client. However, most of the larger companies will have paid by this time. The chances are, if you have not been paid by this time, your client is a small operator, a fly-by-night. At this point, if Small Claims cannot handle your amount, seek a specialist attorney handling collections, since not all attorneys do this. Your local bar association can tell you who handles collections. Be prepared, however, to lose at least 30 to 50 percent of your recovery to the attorney for his fee.

People who sometimes try to get away without paying are small operators or professionals, or one- or two-person business operations, small retailers, and people running for political office. Check out your potential client completely if they are a small business. Often, it is best to ask for business references, or, if you have access to an attorney, ask the attorney for different ways to check the individual out in the community. Do not jump at job offers without doing this.

Q: How do I handle rejections?
A: Rejections are the name of the game. There is no automatic success for 95 percent of filmmakers, no matter who they are, or for most service people, even young medical doctors and lawyers. Do not take rejections personally, but determine to prevail before them.

Work hard to find out *why* you are being rejected. The chances are that it is because you are unknown. Because of this, you are asked to establish yourself thoroughly in your community. This means joining the chamber of commerce and often one or more social clubs. Make yourself known as a volunteer for organizations completely disassociated from your work but politically safe, for example, emergency hot lines, community cleanup patrols, orphanages, halfway houses, reforestation committees, church organizations, or police benevolence associations. The public relations person of your chamber of commerce can help you find such organizations or can direct you to an individual who can help you locate them.

Be careful about publicity when establishing yourself. If you get media coverage, be sure it is for good works, and try to get a line in the coverage as to what you do for a living.

When giving interviews, be extremely guarded about what you disclose. With reporters, there is really no "this is off the record" as you have seen in movies. Reporters look for items that will sell newspapers or increase a radio or TV audience, and the items are not usually good works. Beware of establishing through the media. Try to establish through word of mouth.

To go back to the original question, *always* seek the maturity to field rejections like a "pro." If you are around long enough and are known as a community servant, your rejections will dwindle to a small amount, depending upon other factors.

Finally, be patient.

STRATEGY

The answer to the question about the client not heard from could just as easily have been included among the following tips on how to maintain and build your business once you get started. Following through on your last job should always be a top priority, but there are other ways to keep your business thriving. Consider some of the following tactics.

Take Inventory

Are you expanding your business contacts? Do you belong to any trade organizations? Do you attend trade shows? Search out opportunities to make business contacts. Mix and mingle; after all, this is a business of communication. In all too many cases, it is whom you know that counts.

Devote some time to devising a public relations program. Are you listed in publications, like *The Creative Black Book,* that offer free or inexpensive exposure? Attempt, whenever possible, to place your business name before the public at little or no cost. There are more opportunities than you might think.

Cut Down on Expenses

Whenever and wherever possible, keep your overhead low. If you have equipment that is mostly unused, you are wasting capital. Sell such equipment and rent it back when you need it. Do not let your ego run off with your business. Put cash in the bank for lean times. Too many former filmmakers bought hardware goodies to handle and show off. Do not join them.

What type of car are you driving? An expensive two-seater? Consider this indulgence. If you decide that you can really afford it, hide it from your clients. Many of them will resent it and will not care that it is a result of your success.

Think about all the places you can stop money leaks. Are you having lunch out most of the time? Saving even a dollar a day means $365 at the end of the year.

Organize Your Operation

Plan for the efficient employment of your personnel, if you have any. Teach a less-than-busy secretary/receptionist skills such as mounting slides or canvassing business on the phone. If you work alone, review how you spend your day. Is your day spent efficiently? Are you making new business calls? Schedule them. Opening roads to new business is just as important as calling on existing accounts. It is part of an efficient operation.

Diversify

What type of films are you making? Have you specialized in any particular field? If you have specialized in commercial-spot production, remember that it is a trendy, fashion-conscious, highly political business. If the wave passes you by, you are finished. Protect yourself by diversifying. After you have become established in one field, go after other business whenever you can and, at the very least, keep your eyes open for other opportunities.

Diversification can be quite simple if you consider slide shows or filmstrips. Currently, slides are enjoying a great comeback. More companies than ever before are discovering that training and other communication can be best conveyed by audiovisual technology, thanks to computer graphics. In addition, companies are attracted by the modest budgets involved in using audiovisual communication as an adjunct to their printed programs. The average filmstrip or slide script sells for as high as several thousand dollars, and basic photography (for audiovisual) is billed at several hundred dollars per day; so it is easy to see how your business overhead can be paid by producing one or two slide presentations per month. Take advantage of this trend.

Hybridize

A hybrid production is a combination of film and tape. If your production will end up on one-inch tape or on videocassette, consider transferring your approved film take to tape and completing the postproduction on tape. By carefully planning your editorial time, you can cut the budget by one-half this way. Just compare the cost of completion on tape versus film through wet and dry labs and tape edit. The savings may well be sufficient to purchase a tape deck and some blank cassettes.

Study the Business

To survive and do well in filmmaking, it is essential that you read books and trade magazines to keep abreast of all developments in the film and video industry. Right now, video is in the ascendancy, but it will soon be bypassed by software currently beyond conception. This field is fast moving, and you must be aware of any trends that can affect your business.

Read every trade magazine you can find, paying special attention to the ads. Write to manufacturers for brochures on new equipment. Get on their mailing lists. Do everything you can to learn about new equipment or modifications to existing hardware that can change your business or expand your facility.

After reading this book, if you decide to use it as a handbook in getting started as an independent, return to this chapter to gauge your progress. Once you have grappled with most of the questions asked here and applied the survival tactics to your own business, you can count yourself a success. You will undoubtedly be making money in film.

TWO "BEST HELP" ORGANIZATIONS

It is probably accurate to assume that the Foundation for Independent Video and Film, Inc. (FIVF), and the Association of Independent Video and Filmmakers (AIVF) are, at this writing, the best available assistance to independents. These organizations can either grow or fade away. It all depends upon you, the entrepreneur filmmaker, and your support and participation in this fine process.

Since I believe in it, I solicited and received the following information in a letter from Kathryn Bowser, an official at AIVF and the author of the *AIVF Guide to Film and Video Distribution*:

> The Association of Independent Video and Filmmakers is a national organization providing advocacy and professional services for the independent media com-

munity. Its members include producers, directors, writers, technicians, distributors, exhibitors, media arts centers, libraries, and other individuals and organizations which work in the media field. It provides its members with a national voice by organizing and lobbying around important issues, which recently have included taking the lead in negotiations leading to the establishment of the Independent Television Service, a national funding mechanism for independent projects destined for public broadcasting airtime, representing interests of independents before Congress, the Copyright Office, and with various trade unions.

Through its affiliated Foundation for Independent Video and Film, a not-for-profit service organization, various programs in support of independent media have been established. These include *The Independent Film and Video Monthly*, the only national magazine specifically tailored to the needs of independent producers, covering a broad range of news and analysis, practical information and articles on technical, legal, and business matters.

The FIVF Festival/Distribution Bureau assists U.S. media artists in making informed choices about domestic and foreign film/video festivals; maintains a comprehensive computer database on over 700 festivals; acts as a liaison between 10 to 12 festivals and U.S. producers each year; and publishes guides to international festivals and to film and video distributors.

Other FIVF publications include *The Next Step: Distributing Independent Films and Videos* (Warshawski), *Alternative Visions: Doing It Yourself* (Franco), and guides to film- and video-production resources in Latin America and the Caribbean, Africa, and Asia and the Pacific.

FIVF also runs a monthly seminar and screening program, operates a large mail order book service, and provides information and referrals on production, financing/fundraising, legal and business issues, distribution/exhibition, professional services, educational/training programs, cable, etc.

The FIVF and AIVF are located at 625 Broadway, 9th Floor, New York, NY 10012, (212) 473-3400, FAX (212) 677-8732.

MARKET RESEARCH

The single most important aspect of survival is preparedness by either being well informed or by having an excellent source of information at your fingertips. Market research, when used properly, will help both in planning strategy and in the daily operation of your company—survival.

A company dedicated solely to provide audiovisual, film, and video industry information to professionals is *Hope Reports,* mentioned in an earlier chapter. Here is more detail on the company.

Since this company's business is gathering and selling data, the following information is meant only to give you a small sample of the type of data available. You are urged to secure a complete catalog of this fine company's offerings so you can see for yourself the many ways information can help you survive.

The best of the market research organizations in the industry known to me, *Hope Reports* was founded by Tom Hope, a former Eastman Kodak executive, in Rochester, New York, back in 1970. Through the years, it has provided the industry with an accurate reflection of itself by culling data from trade sources, analyzing it, and then presenting it back to the industry in digested, organized, and viable form. This is tough, grueling work, and it results in an ongoing, annual prognosis of much of the industry as well as numerous individual reports focused on specific topics. It is invaluable information for marketing and planning, and these reports are updated regularly.

Recently, *Hope Reports* expanded its data on film and video production. That resulted in reports on producer operations ("Contract Production for the 90s," 1991), on video post-production ("Video Post-Production," 2 vol., 1990), and on wages paid production personnel and freelancers ("Producer & Video Post Wages and Salaries," 1990).

Hope Reports learned that, in 1990, Los Angeles replaced New York as the number-one media-communication-industry city. AV companies, producers, post-production houses, film labs, dealers, and other service companies numbered 2,250 companies, excluding manufacturers.

"Hollywood is small" when compared to the rest of the film industry (not counting manufacturers of projectors, hardware, etc.). According to *Hope Reports,* "As of 1990, there were some 20,300 companies throughout the U.S. The number of media-related firms grew 21% in just five years."

"Hollywood," or the entertainment industry, has only six major functional companies at this writing: Columbia (division of Sony), Paramount, Warner Brothers, Universal (division of Matsushita), The Walt Disney Company, and Fox. (It is not known if MGM, virtually out of operation at this time, will survive. It probably will.)

The number of video post-production houses jumped 13 percent in the period from 1985 to 1990, while dealer and film-lab numbers shrank. Contract production units reported by *Hope Reports* number 7,000 U.S. companies, with an additional 1,560 in Canada.

The 1990 data showed this U.S. producer breakdown: industrial 3,600; TV spots 1,400; slide/meetings 1,200; and slide service bureaus 800. There are another 6,000 that include proprietary education, training, medical, documentary, Hollywood majors, and independents doing features and television programs, plus many dormant producers.

The most interesting information to those who wrongly believe film has now more or less passed into history, giving way to video, is from Tom Hope: "Kodak sold more 16mm negative film last year than any time in its history. And even a bigger surprise—more money was spent on the slide medium than on video in the industrial world."

Critical in any successful film-and-video production company are the skilled people performing the different jobs. For the first time, a survey of wages paid to various specialists is available from *Hope Reports*; it reveals, for example, that film people receive higher pay than video specialists. The report also gives regional variations in gross revenues and profits. Owner incomes, profits, and perks are analyzed.

Today, with costs rising so much, many producers are reducing their full-time staffs, relying increasingly on freelance personnel. *Hope Reports* wage-and-salary surveys show what freelancers are getting in U.S. and one Canadian city: Atlanta; Boston; Chicago; Dallas; Los Angeles; Miami; New York City; Philadelphia; Rochester, N. Y.; San Francisco; Seattle; Washington, D.C.; and Toronto. Another first is that the report spells out the different formulas used in various cities to determine the pay of writers.

Another area of information concerns the entire presentation audiovisual industry that *Hope Reports* places at $20 billion, which is four times larger than the Hollywood feature-film business. The report "Media Market Trends" spells out the five key markets and principal media, looking ahead to the year 2000. The most valuable item in the report is a table showing the installed base of video and projection equipment used in 20 discrete market segments.

For more information, write to *Hope Reports,* 1600 Lyell Avenue, Rochester, NY 14606, phone: (716) 458-4250, FAX (716) 458-0413.

COURTESY AND GENERAL DEPORTMENT IN SURVIVAL

The late, great Bette Davis put it as succinctly as possible. She said, "On your way up to the top, smile at everyone because you'll see the same faces on your way back down." It goes beyond that, however. For the past several hundred years, business has tended to be part of polite society. Particularly at the top level, people have behaved with manners. Courtesy was the rule, and it always provided a lilt to the day, even those dreaded downer times when everything seemed to go wrong. I know because it was that way when I started in the 1960s.

When you addressed a letter to a chief executive officer (CEO), for example, you always received some kind of response. Not so today, despite the fact that many of today's CEOs get the highest pay ever. Check it out. Write a meaningful proposal letter to the CEOs of the top 25 American corporations, and let me know how many respond. If you get any response at all, it will most likely be from someone who sends you a kind of non-sequitur form letter completely unrelated to your query.

At the beginning of the last decade of the good old twentieth century, many American corporations find themselves in deep trouble. Part of it is because professional attitudes, courtesy, commitment, honesty, the work ethic, camaraderie, and politeness have become unfashionable. Greed, stealing, and short-term quick profits at the expense of all else (including the future) are "in."

If you follow their lead, you are making a tragic mistake. No matter how high you fly, never take your public for granted. Never aim at the level of the lowest fool in the pack. Never take kickbacks or get into any kind of immoral situation. It always leads to a downward spiral—one immoral or outright criminal act always leads to another. Eventually, you might throw away your reputation, your business, and perhaps even your freedom and end up in a cell block. Your survival truly depends upon your honesty and honorable behavior.

Success really does start with how you treat others. The way of life espoused in the age-old dictum "Treat others as you would have others treat you" is always the best approach to business and to life in general. It is not terribly cool or bright to become a cynic, a thief, or a backstabber, for you may eventually end up losing it all, as the following story illustrates.

Back in Rochester, where I spent a dozen years servicing Eastman Kodak and other multinational companies headquartered in that interesting little city of flowers and snow, I met a young filmmaker and saw promise and talent in him. I gave him a good deal of my time and found him to be as charming as he was talented. He was a dashing young man and popular with women, but he soon began breaking a few hearts. My reaction remained on the generous side, giving him the benefit of the doubt.

The years went by and he prevailed after I worked to get him into the industry. He became a top camera-director in the city and then began to get regional, then national, and even international clients. In his wake was a trail of shafted people—colleagues, friends, suppliers, and even clients. The man's business began to slide; and as it did so, the time for revenge arrived. Just as wolves go for the weakest in the herd, the poor guy was attacked by nearly everyone. He eventually disappeared completely from the scene, no where to be found. Besides that, no one cared and no one even asked about him. My reaction? What a waste of talent and life! This man could have survived to enjoy success and a good reputation.

Surviving, however, means more than talent or charm or even hard work. It also means behaving well and being completely aware of others, being considerate every step along the way. When you reach the top, if you are that fortunate, then is the time to truly watch your behavior. It is then that all eyes will be cast in your direction.

SURVIVING THROUGH PERSEVERANCE

My favorite story of perseverance comes from old Hollywood, and it is the life of Carl Laemmle, one of the most interesting of the early pioneers. Ephraim Katz, in his truly extraordinary 1979 publication *The Film Encyclopedia,* has a biography on Carl Laemmle, which is excerpted here with permission from its publisher, HarperCollins Publishers, Inc., New York, NY. © 1979 by Ephraim Katz and See Hear Productions.

> LAEMMLE, CARL. Motion picture tycoon. *b.* Jan. 17, 1867, Laupheim, Germany. *d.* 1939. The tenth of 13 children of a middle-class Jewish family, he became gainfully employed at age 13 and was an experienced bookkeeper and office manager at 17 when he decided to seek greater opportunities in the New World. Arriv-

ing in America in 1884, he became an errand boy for a New York drugstore, then went to Chicago, where he held a variety of jobs, from newspaper delivery to bookkeeping. Moving on to Oshkosh, Wisconsin, he worked his way up to clothing store manager and married the boss's daughter. Returning to Chicago, he decided to invest his savings in a nickelodeon. He opened one in January of 1906 and, encouraged by the quick returns on his investment, launched another within two months. Unsatisfied with the quality of service by the local film exchange, he set up his own exchange, the Laemmle Film Service, in 1907. He subsequently opened exchanges in several other American and some Canadian cities and before long was among the leading distributors in the business.

In 1909 Laemmle courageously defied the pressured tactics of the Motion Picture Patents Company, which had put many other exchanges out of business. Not only did Laemmle refuse to sell or fold his company, but he also announced that he was going into production in direct competition with the trust. He promptly founded the Independent Motion Picture Company of America (known in the business as IMP) and released its first production, *Hiawatha*. He began an enormous publicity campaign to discredit the Patents Company and build an image for his own studio.

In 1910 he pulled the clever publicity stunt that launched the star system in American cinema. Having lured from Biograph its most popular player, Florence Lawrence (she was known then only as "The Biograph Girl," since Patents Company producers feared that naming their players would result in the upping of their salary demands), he planted a report in the newspapers indicating that "The Biograph Girl" had been killed in a streetcar accident. The following day he came out with an indignant advertisement denouncing the malicious report and announcing that Miss Lawrence, now "The Imp Girl," was alive and well and working for him. He also lured away Mary Pickford, announcing that "Little Mary is an Imp now."

Intent on glorifying his company's product (100 shorts by 1910) Laemmle glamorized his players and spent unprecedented amounts on publicity, regularly mentioning his star by name. Meanwhile, he successfully blocked continual attempts by the Patents Company to put him out of business and in 1912 won a court battle that hastened the demise of the trust. In 1912 also, IMP merged with several smaller companies to form the Universal Film Manufacturing Company, a major studio (later known simply as Universal). In another pioneering coup, Laemmle's company turned out a feature length expose of white slavery, *Traffic in Souls* (1913), which proved to the industry that there was money to be made in feature films and in the exploitation of sensational subjects. The film cost a little over $5,000 and grossed close to half a million dollars.

The diminutive 5 ft. 2 in. and notoriously eccentric Laemmle celebrated the high point of his career in 1915, when a crowd of 20,000 gathered to watch him officiate at the opening of Universal City, a 230-acre studio municipality in San Fernando Valley. Many famous stars, directors, and executives started their careers as Laemmle employees. Irving Thalberg and future magnate Harry Cohn began as his personal secretaries. But the first allegiance of "Uncle Carl," as he was amiably called in Hollywood, was to members of his own family. He gave the key job of Universal's production chief to his son, Carl Laemmle, Jr., as soon as the latter turned 21, and placed 70 other relatives on his payroll in various capacities. Laemmle Jr.'s extravagances during the Depression years compounded the financial difficulties the company was experiencing at the time. In 1935, Laemmle Sr. was forced to sell Universal for a little over $5 million. He lived just long enough to see his former company bouncing back to health thanks to a singing teen-ager, Deanna Durbin.

SURVIVAL THROUGH OBJECTIVITY

Laemmle's biography should carry several distinct messages for you:

1. It does not take a tall person to make it, just one who can place fear aside, set one target after another, and go for it. Note how Laemmle was never frightened (while others were), and he was all of 5 feet 2 inches.

2. If the established powers got there, you can get there. All you need is imagination and perseverance.
3. No matter how mighty you are, you can still have an Achilles heel that might cause your downfall. Laemmle's was nepotism, an objective blindness.

A similar story is told by tiny Japan. At the beginning of the century, it was spreading through parts of Asia going for conquest and an empire. In 1945 it suffered a stunning defeat that nearly destroyed its society and altered its very culture. It then struck out once more to create an economic empire, with great success, eventually acquiring even Universal nearly a half century after Laemmle had to sell it. Laemmle would be delighted. He was that kind of man.

Nothing will stop Japan's forward motion except, if it exists, an Achilles heel. People, and nations, prevail until something *within* stops them. It has always been so.

In summation, survival is achieved not just by being patient, long-suffering, and persistent, but by willing to be inspired by the lives of others who have walked the same path before. You survive by continuing your work when there is no external work order. How do you do this? Pursue as much industry information, data, and connections for networking as possible.

In lecturing to groups (most recently a few days ago, as I write this, at Brooklyn College), what I often tell them is to become *political.* If you do not have a clear definition of the word, look it up, and learn to live with it until it is completely internalized and integrated within yourself.

My final word on survival must be *inspiration.* As you have seen, others can inspire you. Others, however, cannot truly motivate you. You must do that yourself. How can you do that if you do not have a role model at home or in school to show the way? You can turn to professional motivators. The very best among all the ones I know about is Nightingale-Conant, a company I cannot say enough good things about. Working with their many excellent audio programs on cassette can completely transform you or help you in one of the weaker aspects of your personality, if you are already in the Superman/Wonderwoman category. Nightingale-Conant's toll-free number is 1-800-323-5552. Ask for their catalog.

Yes, there are many others who can and will help you survive; but the only one upon whom you must utterly depend when the chips are down is yourself, not lady luck.

7

PROMOTING YOUR BUSINESS

The young woman had large blue eyes and hair the color of unpicked corn in midsummer. She was Scandinavian and appeared to wear no makeup, just false eyelashes. As she moved in front of the camera, the crew worked to place light to make her hair gleam and enhance her features with softness.

Everything seemed to be going beautifully. Looking across the room, I caught the eye of the producer. He smiled and winked at me as if to say, "Everything is going very well. Hope you are okay." We were just about to shoot some expensive 35mm footage.

The producer was an old friend who had hired me to direct for him on other projects in the past. I signaled him to wait, walked over to him, and asked in a hushed voice, "Where is your still man?" He looked around and then down at the floor. "Uh, we have no budget for a still photographer . . . "

"But this is a new product," I said, carefully forming my words. "Didn't anyone from the client suggest still coverage?" He looked at me and again his eyes sought the floor as he said, "No, but I should have."

In overlooking the need for a still photographer, my friend failed not only his client but also himself. It will become clear as you read this chapter that he missed an opportunity to derive promotional material for his own business while extending the scope of his work for a client. The filmmaker who is also a canny businessperson will always be looking for such opportunities.

One of the keys to your longevity as an independent filmmaker will be your ability to promote and merchandise your services successfully and creatively. Several effective means of doing just that are the basis of this chapter. Although these tried-and-true methods can be valuable in locating new business, they are just a starting point: Only your variations will make a promotion work for you. Furthermore, some of the advice herein will sound familiar because it is an elaboration on the business basics discussed earlier. Clearly, practical application of the basics will be at the center of your success, initially forming the building blocks of your fledgling business and later the foundation upon which creative promotion is based.

HOW TO MERCHANDISE YOUR BUSINESS

Filmmakers presenting their services to prospective clients must create in each prospect a desire to see their wares. Prospects must understand that you have studied their communication problems carefully and that you are genuinely interested in their welfare. After all, the success of a client's business will mean success for yours as well. By carefully planning your presentation, you can nurture your prospect's interest in your business.

Before you think of contacting a prospect or an old client about applying a filmic solution to a communication problem, be ready to reply to all objections. That means devising three answers to each anticipated problem. Since the usual objection is budgetary, be prepared to show your prospect how the extended use of film compares favorably in cost to another medium. The key, as always, is preparation. The following is a review of some of the basic steps from the perspective of merchandising.

Know Your Prospective Client

Who is your prospect as an individual? Study your prospect's habits, preferences, and dislikes. Would the prospect consider a hard sales pitch offensive? If you do not know, take the individual to lunch and find out. Do not be afraid to ask questions that touch on personal habits, although you should take care to place them in context. Always use tact and diplomacy. Every person has a different level of tolerance for probing questions, so try to communicate sensitively. Move the conversation away quickly from the weather and Saturday's ballgame and talk about business. Above all, however, listen to your prospect.

Once you know your prospect, you can cater to the individual's attitudes and preferences while merchandising your services. Point out that motion pictures, video, and audiovisual presentations have an extra dimension that results in recognition for those using these media. Drop an occasional hint about the visibility resulting from screened presentations. Stress the cost effectiveness and memory retention produced by the multisensory impact of film. This is merchandising on the personal level.

Know Your Prospective Client's Company

Who are the employees? How is management arranged? What is the business history of the company? Who is their competition and how do they compare with the rest of the industry? How well do they communicate with their public? Have they commissioned a film production before? By whom? You cannot merchandise the company until you know the facts.

Determine whether your prospect's competitors have used film. If they have, point out the necessity of matching competitors in the marketplace. On the other hand, if the competition has not adopted film, argue the advantages of beating them to it. The amount of interest you generate about your work will infect the potential client and, thus, the company. If you lack enthusiasm, so will your prospect.

Plan a Good Presentation

After you feel you know enough about your potential client to present your ideas, start planning. Remember to design the presentation around what you know about your prospect's needs. Think in terms of true benefits for your prospect, who is bound to ask, "What can you do for me?" You must provide a satisfactory answer if you want to be hired to produce a film.

Keep your presentations short, but be sure to present all your points. If you have made a film that relates to your prospect's needs, show just the first five minutes or so. Then turn off the projector and briefly describe the project's benefits. (If you are asked to show the rest of the film, great—but do not be crushed if a prospect cannot spare the time.) If a sales promotion film you produced enjoyed wide distribution, you should be able to acquire a history of the product's performance before, during, and after distribution. Present such findings to your client on paper. Quite often a satisfied client will be pleased to write a testimonial on your work. Do not be timid about soliciting such recommendations if they are deserved. Then use them to strengthen your presentations.

When you present a reel or cassette of television spots, use only your most exceptional and recent spots. Edit in only those spots that relate to the product or service family in which your potential client deals. If you have a particularly impressive spot—a big-budget job or a Clio winner—try to determine its effectiveness. Solicit numbers from the market research people or advertising agency for whom you made the spot. Was there any measurable sales increase after the campaign? How did the spot pay off for the client? Clios are wonderful for your ego, but they do not impress smart clients; so emphasize hard sales data, not awards. Merchandising means talking performance.

Follow up presentations with a call or letter of thanks. A show of gratitude will go a long way to indicate your interest and sincerity. Potential clients do not like to feel they are being hustled. If you fail to win a job after a presentation, remain in touch anyway. Let your prospect know that you are available, ready, willing, and able to help with communications problems that can be solved with film. Do not hesitate to offer to screen another film if you think the prospect would benefit from seeing it. Such follow-through can go a long way in cementing a future business relationship.

DIRECT MAIL THE 3-D WAY

Canvassing is not the only method of getting your foot in a prospect's door; a direct-mail campaign can be equally as effective. To be successful, however, your prospect must read your mail and react to its message. Therefore, consider the power of 3-D campaigns.

Be honest. How much attention do you give to a letter hyping someone's business, even when it is expertly or cleverly written? Chances are such mail lands unread in your wastebasket. 3-D is something that comes in a box. It engages the receiver's curiosity.

Earlier discussions of direct mail stressed the need for color or an unusual graphic technique to arouse attention. If you can afford 3-

D, however, it can even more substantially improve your recognition in the market. It also makes sense for a filmmaker to make the strongest visual statement possible.

There is some good flat mail that has character, pizzazz, or some utility. Consider the Fotomation film timetable or the small slating card sent by photoanimator Al Stahl of New York City. The timetable has a gauge/film footage scale lined up with seconds, minutes, and number of words per foot. I use these gadgets often and guard them with my life. Someday I will use Al's animation camera again because he is so clever with his direct mail and always delivers the goods. That is the crucial point here. If you are going to invest in direct mail—whether 3-D or flat—be prepared to produce results.

To illustrate the power of 3-D, consider an extremely successful mail campaign devised by the Marsteller advertising agency for Clark Equipment Company. Marsteller called it, "the case of the two-pound calling card," and it ran as follows.

In 1964, six-year-old Clark equipment was the country's fourth largest supplier of trailers. Their problem was that 300 major buyers did not buy from Clark and never had. The direct-mail campaign was designed to establish direct personal contact with these nonbuyers.

The campaign began by sending off a chef's hat with this message: "Clark salesman wear many hats—equipment specialist, financier, service man. A good reason to buy Clark trailers." A week later, three gourmet-style barbecue skewers were dispatched bearing the story that Clark had the right equipment to handle the customer's job.

Message three—also sent after a week—comprised salt and pepper shakers indicating that Clark had the extras that are important to trailer buyers. The fourth week, a cookbook was mailed with a message about the firm's wide experience in matching products with customer needs. The next gift, a padded glove and apron, symbolized downtime protection. Clark had that, too: a nationwide service setup.

Each package and message was accompanied by a salesman's business card that said "stand by." During the sixth week of the campaign, a two-pound prime filet steak was delivered by the salesman in person. The message: "This is the kind of personal service you always get from Clark." Personal contact was established with 300 prospects and resulted in 50 new accounts. An investment of $12,000 had yielded $3.5 million in new sales.

This campaign illustrates the effectiveness of 3-D direct mail. In order for this concept to work, discipline is demanded on several levels: discipline on the client's part to reduce the prospect list to the smallest number possible, usually a few hundred; creative discipline to insure that individual pieces never become gimmicks and overpower the sales message; finally, the discipline to make follow-up research an integral part of each 3-D campaign to determine the return on the communications investment.

The average filmmaker soliciting new clients does not have the resources to mount elaborate 3-D campaigns, but studying such efforts can suggest some successful approaches. What can you afford in terms of 3-D? Start by checking out an advertising specialties company. Their staples are pens, desk accessories, and other utilitarian objects that carry a company name. Some creative houses will collaborate with a customer in devising a unique item that communicates and can be produced at a low cost. A 3-D idea can be found in a novelty shop or bookstore and purchased in volume at a reduced price. Just remember the item must be tied in to your pitch. If it is useful, so much the better. The success of Al Stahl's timetable is that it provides me with a reference that I often need while reminding me of his name each time I use it.

When selecting a 3-D item for a direct-mail campaign, always think of your prospect's needs, their businesses, what they do, and how they do it. Start with a list of all your business contacts; then trim the list by qualifying each prospect for the campaign using the folowing criteria:

1. Has the prospect used film for communications before? How? For sales promotion? For training?
2. Does your prospect have a long-term contract with one of your competitors?
3. Have you read your prospect's annual report to help you determine its readiness for film?
4. Is the prospect in a position to sponsor a film? Is the company "film wise"? Have you seen any of its past productions?

Once you have your list of qualified prospects in hand, the items and message for your 3-D campaign may be easier to pinpoint. Direct mail, like most other ventures, is not worth doing if not done well. Do you have the correct name and title of the person responsible for the selection of film-production services? the proper address? Make sure your pieces of direct mail are seen by the right people, otherwise you are wasting your time and money. Proper preparation and planning are the key to a successful campaign that can establish firm business contacts and the corresponding opportunities to make a sale.

STORYBOARDS AND PROMOTION

One item that filmmakers find successful as part of a direct-mail campaign is a storyboard depicting their film or television spot. It is an effective means of using your current work to promote your business and generate future jobs.

A storyboard contains several picture frames, usually one for each scene in your spot or film, accompanied by appropriate captions. The front of the storyboard can also carry a sales message or mention unusual methods used in production. Where desirable, the back can be filled with additional text promoting your production facility and any special features you may offer, for example, mixing or special effects. The amount of copy is discretionary, but keep in mind that the advertising industry is built on short, to-the-point messages.

Before initiating production of a color-print storyboard, obtain written permission for the use of your client's film or spot in your sales-promotion efforts. Regardless of whether you have chosen as few as six frames or as many as twenty, you are still using material for which your client has paid; therefore, the client owns the rights. Take this opportunity to offer your client a quantity of the storyboards. These can be either sold to the client or presented gratis as a goodwill offering. If you approach your client to share printing expenses for the storyboards, explain the various functions they can serve: point-of-purchase (POP) displays, new-product introductions, sales catalogues, dealer promotions, and direct-mail pieces.

When you are shooting 35mm, full-color boards can be printed directly from key frames taken from one of your quality prints. Assure yourself of high quality by working from a print that has never been projected. If you are working on videotape, any number of frames can be shot directly off a good television monitor; 16mm can also be copied from a video screen, but you will get far better results by employing a still photographer on the set while you shoot in 16mm. Direct the photographer to cover the subject at or very near the motion picture camera, using the same angle and lens exposure so the resulting shots have the appearance of the subject as shown on film. For animated spots, storyboards are best produced from 35mm stills of key frames taken directly from the cels.

When selecting the frames to be used, keep the audio in mind. If you can fit matching audio under each frame (which I recommend) and have space enough for one frame for each scene in the spot, you have planned well. Generally, the amount of audio to be used as copy plus the number of scene changes determines the number of frames in your storyboard.

When you are ready to produce your printed storyboard, select a printing house experienced in this type of work. Not only is the average printer not equipped for such specialized work, he or she is also likely to charge more than specialty houses. Visual Promotions, Inc., one of the largest and best known printers of these colorful merchandising aids, even offers a kit of instructions for the correct preparation of the board that will help save additional cash. Generally, the printing house will need a rough layout, typed copy, and strips of 35mm film. Discuss the job with your printer in advance to insure that you have supplied everything that is needed for the job.

Once you have the boards, do not limit their use to direct-mail campaigns—carry them on sales calls; frame them and hang them on your waiting-room and office walls; hand them out after screening a sample reel. The storyboard style is flash. It looks slick and tells the story in pictures. Since your audience is composed of clients or prospects looking for a visually sophisticated studio, storyboards provide an excellent spotlight on your talents as a visual communicator.

YOUR OWN PUBLIC RELATIONS

There are several options other than storyboards that provide equal opportunity to boost simultaneously your own business and that of your client. Such methods are public relations just as surely as those discussed in Chapter 5, only here the emphasis is broadened to include media outside film and video, particularly print.

Whenever a production involves a product or service that is newsworthy, it is advantageous for both producer and client to exploit their work. In most cases, if there is a news angle, it is worth publishing. The type of news angle will dictate where it is published. For example, if the film covers a new product that will affect everyday life—such as the first demonstration of a videodisc—the news angle suits the mass consumer media. Information of interest to manufacturers only, on the other hand, would be better targeted at trade publications, such as *Women's Wear Daily* or *Steel World*.

Remember, nothing angers editors more than receiving a news release that has no news value and is an obvious plug. Most are willing to read news releases from recognizable sources, particularly those known to have integrity. The trick is maintaining enough objectivity about your own work to judge its value to the public. Training films, for example, are not ordinarily considered worthy of exploiting in any medium unless there is a news angle. What kind of news angle? Well, does the film utilize a new form of training methodology? Are the trainers newsmakers? Is the filmmaker a newsmaker? If Steven Spielberg made a training film, it would certainly be considered news.

Many magazines have editorial guidelines and it is wise to write to the editor for these before sending off news releases of feature articles. Once you have become acquainted with the editorial policy, formats, and slant, you can feel confident that features written for specific client objectives will conform. Also, answer the following questions for yourself before attempting to write any release or story:

1. What are the publications in this market?
2. How much space are they likely to give this story?
3. What is their level of information regarding this product or service?
4. What types of "newspeg" have the greatest appeal in this market?
5. How much rewriting or release reading time are they willing to invest in this subject?
6. What type of information is generally acceptable in this market?

Following such a review, you should be able to determine whether a release or a feature story will best serve your purpose. For a feature, query the magazine about their interest: send off a one-page description of the story accompanied by suitable illustrations. Include a self-addressed, stamped envelope for return if the editor is uninterested. If you receive a go-ahead, it will most likely specify the article length in words, photo sizes, and instructions on specialized interview formats. Follow the editor's instructions, if any, for manuscript submission; otherwise, submit cleanly typed copy, double-spaced with 1½-inch margins all around.

If you decide that a feature article is needed and you are not a writer, do not hesitate to hire one. Public relations for yourself and your client are important to your success, so it is worth it to invest in professional help. If you enjoy writing, you may wish to move beyond writing for clients and submit technical articles to film trade journals.

You may decide that a news or press release would be more suitable. In this format, you will provide extensive information in the most concise manner. Generally, a press release touting a new product or service would include the following features: an announcement of the product's development in a summary sentence; a definition of the product and its significance; an explanation of how it works and its applications; additional product features; advantages for the consumer; documentation of claims; specifications; availability; client puff. The sample release that follows, which was prepared for a construction trade magazine, contains most of these components. Can you identify them?

VIBRO DRIVER TRIMS JOB SCHEDULE FOR STEEL PLANT SETTLING BASINS

A vibratory driver/extractor drove pile sheets in three to four minutes per pair during construction of a new waste handling system for U.S. Steel Corporation's National Works, McKeesport, Pennsylvania. Speed of the vibratory unit cut pile driving time and costs, and permitted the job to be completed ahead of schedule, according to the contractor, Dravo Corporation.

Dravo used a Model 2–50 Vibro Driver-Extractor, purchased from L.B. Foster Company, to drive pairs of MZ27 and MZ38 steel sheet piles. The piles form sidewalls for three settling basins at National Works. Driving rate was 121 feet per hour at the start and 230 feet per hour at completion, averaging 171 feet per hour for the job.

Pile sheets were driven to refusal through Monongahela River silt and clay. The Vibro required no downtime. The only maintenance required was lubrication between shifts.

Dravo used two steam hammers—a 9B–3 and a 10B–3—to tack wall sections of piling before driving. Actual driving time for the Vibro was three to four minutes for a pair of sheets. There was no deformation from driving. The piling was cropped one foot from the top and capped with a 15-inch channel section.

The heavier sheet piling was used on the outside walls of the basins to withstand buffeting forces of the river. Because of barge activity, the outside wall of the second basin was reinforced with pile clusters. Prefabricated waler sets were barged to the job site from Dravo's Neville Island plant downriver. They were used as templates during pile driving.

The inside frameworks of the basins consisted of H-beam walers supported with horizontal pipe bracing filled with concrete. Wingwalls were added to the upstream end of each basin for protection against flotsam.

> The settling basins will contain industrial wastes from the pipe, bloom, and bar mill areas at National Works. Mill scale will be collected in each basin where it will settle for periodic removal. Oil will be skimmed at the top while the remaining clean water is returned to the river. U.S. Steel's central engineering department designed the new waste handling system.

Note that client puffs must be light and woven into the story. The reader does not want to confront an advertisement in the guise of a press release, and editors are severe when confronted with too much hype.

In addition to writing releases, there is also a role for photography in public relations. Quite often a magazine will publish a picture that says quite a lot without hanging a story on it. Consider a photograph of a commercial being filmed on the Avenue of the Americas in New York City with Jack Lemon as narrator. He wears the uniform of a Minuteman; a crowd of smiling people stands behind him. Little needs to be said except to identify the scene in a caption. Readers read pictures, as every photographer knows. This is called visual literacy.

MODEL AND PERFORMER'S RELEASE

Subject ______________________________

Project # ______________ Hrs. ______________

For value received and without further consideration, I hereby consent that all photographs taken of me and/or recordings made of my voice and/or written extractions, in whole or in part, of such recording or musical performance at ______________ on ____________ 19 ____ by ______________ for (Company) may be used by (Company) and/or others with its consent, for the purposes of illustration, advertising, or publication in any manner.

SUBJECT:______________________________
(signature)

Street ______________________________

City ______________ State ______ Zip ______

IF SUBJECT IS A MINOR UNDER LAWS OF STATE WHERE MODELING IS PERFORMED:

GUARDIAN ______________________________
(signature)

Street ______________________________

City ______________ State ______ Zip ______

Most of the time, however, photographs are used to illustrate a story. They help create lively layouts in otherwise stiff trade magazines. They also complement the story often by clarifying or amplifying what the writer is trying to say.

Whenever you provide a photograph for publication, be certain you have a signed model release for each person portrayed. A release should be obtained whenever you are shooting a film; you can never predict how you will publicize your work or when a performer will seek to assert individual rights or protect privacy. Legal model-release forms are available from film equipment suppliers and should be used on every production. (See example on page 93.) Note that a parental signature must be secured if you are using models who are legally under age.

Whether you use direct mail, storyboards, photographs, or feature stories, you cannot succeed unless you become promotion conscious. As you work on a day-to-day basis, stay alert to all promotional possibilities. For public relations, think in terms of news angle for any kind of exploitation, exposure, or information that you can pass on to the public in a way beneficial to you and your company. At the same time, remember that any public relations must have validity in the medium you are approaching. A newspaper is usually interested in fastbreaking news stories. Most magazines lack the urgency of dailies but require stories that interest their readers. Television demands flash-type news before it provides coverage, although the proliferating television magazine programs offer new possibilities for film public relations. The key is to be alert to the opening and always be prepared.

8
CABLE TV AND OTHER COMMERCIALS

In the late 1960s, at Clarke's Bar on Third Avenue in New York City, I met with two friends whom I had not seen since college. Clarke's is a wonderful old advertising pub that, for reasons I could never fathom, is mistaken by everyone I have ever met for P.J. Clarke's (which it never was). It has been saved from demolition and now sits happily on the corner lot of a sterile, modern skyscraper, a charming reminder of early twentieth-century New York City when the bottom line wasn't the end-all of life.

I was in town from the Pittsburgh branch working on a project for the large public relations firm Burson-Marsteller, which still had its home office on Third Avenue, not far from Clarke's. My two friends were account executives for two of the big agencies of the period, Ogilvy & Mather and Young & Rubicam, which are still big today.

Being advertising people, they sort of looked down on me, a public relations flack. I was just then getting back into film and audiovisuals, which I was introducing to Burson-Marsteller. They were strictly a print media firm in those days. My two friends poked fun at me, actually, and I laughed it up with them, something we all do when we are young and stupid. Later, we only have time to complain.

We speculated as to how broadcasting would change advertising. Cable TV was so new it never entered the conversation except as a one-line gag. But highly biased by my own work, I wanted to tell them how receptive I felt was Burson-Marsteller to audiovisual technology and to the many new ways I could envision public relations problems being solved by film, AV, or the new videotape recording, which was then still black-and-white in industrial use and in color for only a few years in broadcasting.

This was the beginning of the big era of multiscreen presentations, and I saw my future as bright as the Bermuda skies. Unfortunately for me, however, I left Burson-Marsteller just a few weeks later for a salary offer, a doubling of my paycheck, which, as a newlywed with my wife still in college, I could not turn down. The agency I went with, a small upstate New York firm that had promised a big audiovisual future (they handled Eastman Kodak's industrial advertising, so what could look better?), never really went anywhere with AV, lacking good top management at that time. In the meantime, what I started at Burson-Marsteller later became a major division of that agency.

I should have taken the time to thoroughly investigate both firms instead of making a quick jump for money and false promises. If I could only have looked into the future!

FORECAST: AGAIN, LOOK FOR CHANGE

As of this writing, the FCC has granted usage of telephone lines as television-signal carriers. This is possible because of two parallel technologies that entered about the same time and have been put into use to change the way communications are carried via cable: fiber optics and digitization. Before this, phone lines could carry only a primitive still picture.

What this will do to existing cable companies and how far it will go, no one knows. Anything can happen, especially in communications, which has changed radically in ten years and is changing the world.

Cable advertising represents a comparatively new field where everything is up for grabs. Since it is wide open at this time, there is no agenda, no crystallized time structure, and no budget parameters to discuss.

Cable television itself is still growing, thus, still structuring itself, with no formats. This works for the benefit of the video-filmmaker. The first commercials on CTV were taken directly from network and independent TV, changing from 60-second or longer to as short as ten-second spots.

The past few years have seen the growth of "commercial" programs in which the entire program is just one, long, sometimes quite interesting, sometimes deadly commercial. This type of programming had its origins in early TV but has now come back with a vengeance in CTV, so much so that it can now also be seen in independent TV during "off hours."

The budgets for commercial programming are all over the ballpark. In all of TV, the content and the quality of the pitch is what sells, not how little or how much is spent on production. This has always been the case, although some wily spot producers have convinced gullible clients otherwise. Woe be it to producers whose clients finally get the word after being had.

During the past two decades alone, of commercials on network TV, budgets have multiplied tenfold. Campaigns costing millions of dollars have sometimes culminated in an as-yet unsold product being taken off the shelf. Expensive "clever" or "cute" commercials have resulted in zero sales. All this is because many clients forget what a commercial really is. It is a pitch, and nothing more. Slick commercials that sell nothing are parasitical and wasteful.

PITCHING IS THE SELL

Selling is the focus, the target for all pitchmakers. When commercials are made, they are supposed to be selling something to someone, not parading "creative" ability at the expense of the consumer (who ends up paying for all advertising). Ultimately, the client's bottom line will be what determines a pitchmaker's future, not awards for "creativity."

The most disastrous TV spot campaign in memory was the introduction to the new Nissan luxury line of autos under the "Infiniti" logo. These commercials showed forests, shorelines, birds, and so forth, and spoke in vague poetic language. Not a single product was shown, so no one knew what the commercials were nor what it was that was being sold. An astonishing waste of money, this went on for many months, irritating the prospective buyers instead of selling them, while the Toyota luxury line, Lexus, coincidentally introduced at about the same time, took a great lead in the market place because it was selling cars, not some "creative" advertiser's strange conceit.

It is a wonder that the advertiser got the account in the first place. I envision the Japanese Nissan liason now working in some northern island rice field.

To date, Nissan has not yet recovered from that campaign, one that became the brunt of jokes and embarrassment all through the advertising industry. It was obvious that the advertising agency had no idea how to sell cars.

Ironically, there is a lot to sell in the new luxury automobiles from Japan. They are examples of extraordinary engineering and fine quality-conscious production. Decades ago, Japanese industrialists, perhaps before most of the rest of the world, realized that in the new world order of manufacturing, with the new global economy and consciousness, quality had become mandatory for survival. They began to build high-quality products and to support them with good service and maintenance. Yet, the American marketing team and the advertising agency, apparently oblivious to this new understanding, were missing the entire picture, undoubtedly, a failure in communication from top to bottom. The single most prevalent cause of business failure is a breakdown in communication, a glaring fact ignored by many companies.

The most successful merchandisers of automobiles in the luxury line have been BMW and Mercedes and other German-manufactured cars, who sell, quite directly, engineering and safety. Note that German auto manufacturers do not identify their cars by animal or bird names, nor do they try to dazzle consumers with fancy footwork to sell "the sizzle." They simply label their autos with engineering designators and sell engineering. They know they have something to sell because they have worked hard at creating the products as functional, practical, reliable vehicles. Even if they sometimes look like less than gorgeous Italian sports cars or jet aircraft, nothing made yet compares to their engineering. Most people who can spend the money to buy a German car are not fools. They buy hardware that performs well, not poetry.

A NEW GLOBAL AUSTERITY

With the new austerity of CTV, independent TV, and whatever else is coming down the line, selling will once more take precedence over games of shadow and illusion.

The greed, thievery, and irresponsible fiscal management of government of the 1980s plunged the United States into a prolonged recession bordering on depression, with record numbers of people on food stamps and unemployment lines. As in the great depression following the stock market crash of 1929, which was the climax of similar financial craziness following World War I, people will have a long memory of the hard times, and recovery will be slow.

For filmmakers, the outlook is not bad, however, for communications is still balancing out and spreading. The opportuinities are there for those who seek them.

THE UPLINK CONNECTION

New and existing companies, those with some CTV or other experience, are now creating programs entirely for free access to satellite uplinks in order to get products sold. This is much like "free" advertising network development in the 1940s when CBS, NBC, and, eventually, ABC were formed. In this field, however, it is much like sprinkling global seeds, for there is virtually no limit to this new field of technology and no one has a clue how far it will go.

As always, it is best to scan continuously the reference books in the local library for breaking forms and technologies such as these. The *International Motion Picture and Television Almanacs,* issued every year by Quigley Publishing Company, Inc., 153 West 53rd Street, New York, NY 10019, have been particularly helpful to many in the business.

LOCATING PROSPECTS IN CABLE AND OTHER TELEVISION

The basic front door is the advertising agency. Each advertising agency is quite different from the other. This is so even when traveling to different-city branches within a single agency, such as BBDO. Each manager has a different style and hires people with varied backgrounds; thus, the office becomes a projection of the manager.

The basic rule follows, however, that the creative team hires the film- and videomakers. One can ask to see the creative director (in larger agencies there is an entire hierarchy of individuals under that nomenclature, such as executive creative director and even senior executive creative director). This title did not even exist, for the most part, about 30 years ago. There were art directors, copy chiefs, account management. However, there is now a particular creative director (CD) assigned to a particular account, depending upon the size of the agency. In some cases, the CD is not the person to see, especially if there is an executive producer in the agency pecking order. In smaller agencies, especially when one already knows the particular account or is somehow associated with personnel at the account, a filmmaker can call on the head account executive. This title will more or less take the form senior vice president, account head, or account group supervisor.

When not certain, try talking to the receptionist. An actual visit is far better than a phone call. The receptionist is not one you need an appointment to see, so just walk in and start asking questions. Unfortunately, some ad agencies, which are supposed to be treasure houses of communications technology and expertise, do not communicate well within, so do not be shocked if the receptionist knows nothing and no one.

Once you locate the correct individual, and it is always best to double check this with at least two sources, then it is *always* best to call that person, not write. Letters and even faxes have a way of finding the circular file quickly. (When faxes were new, they somehow seemed more important than letters; so they were heeded. This novelty faded rather quickly and today they are considered nuisance and, to a great degree, clutter.) A phone call pressing for a meeting (as discussed in earlier chapters) will net you entrance far more quickly than a letter or a fax. Always ask, "Who hires film or TV producers?" or "Who is responsible for film and television production?" The direct, clear inquiry leaves little chance for misunderstanding.

In Syracuse, New York, when you walk into an agency called Mower Associates (its name at this writing), you will see a slogan on the wall selling the attributes of "persistence." Although I disagree with this as a general rule, there is something to this approach. Persistence, however, must be balanced off with clever strategy.

One way to improve your chances for success is to precede your phone call with a brochure, flier, or even a PR article you might have placed in the local newspaper covering some of your work, as discussed in Chapter 6, Survival. In this case, a letter becomes more important and, as "cover letter," will have a better chance of being read.

The letter must be informative. See the example on page 101 for a good example of such a cover letter. Note that the numbers at the extreme left of the letter are *only* for review here and not to be used in a letter of any kind. Here are points to remember when writing a cover letter:

1. Always have a letterhead. This should be a professional looking, aesthetic, businesslike letterhead. Clients, particularly industrial clients, are not impressed with funky or cute letterheads. To them, this denotes a lack of business demeanor. A letterhead can

1 YOUR LETTERHEAD
today's date

2 Ms. Diane A. Smith
Executive Director
JONES ADVERTISING, INC.
235 Main Street (Suite 123)
Big City, USA 00000-0000

3 Dear Ms. Smith:

4 Enclosed is an article you might have seen in the *Times Journal* that ran in the business page on May 21st of this year.

5 The article covers our production for an ABC Kitchen Magic Mfg. TV spot we successfully completed on schedule and under budget. Our research following the broadcasting of the commercial revealed a coincidental rise in sales of 48 percent over the period prior to advertising. This was one of half a dozen such productions, all of which had similar results.

6 We take great pride in our service and have worked very closely with all our clients, one of which is Thomas Harding, with whom I understand you once worked.

7 Within the next week, I will phone you to set a time for a brief meeting in order that I may provide some additional background and data for you.

Sincerely,

8 Your name
Your title

9 enclosure:

---

help or hinder your chances for success. Always place today's date in a prominent place at the top of your letter, usually centered or on the right side.

2. Always be sure you spell the name correctly. Double-check this. Always be sure you have all the other data correct, including the person's exact title, and all numbers. If the agency is housed in a multifloored building, you must include the suite or floor number. Always include, also, the entire zip code.

3. Always address your client (prospect) formally until you have established a relationship. Do not write to a stranger using just a first name until you have earned that right to do so. Today, any woman is correctly addressed as Ms., not Miss or Mrs.
4. The introductory paragraph is brief and states the reason for the letter.
5. The second paragraph goes into the reason in detail and offers additional information on yourself.
6. The third paragraph supports and enhances the second, then ties you in with a business connection or reference known to your prospect.
7. The fourth paragraph sets up action for a meeting in a positive, affirmative way. Note that you are also saying the meeting will be brief but beneficial to your prospect. When writing any kind of letter, try to see it from the reader's point of view. How will it be received? How would you react to any letter you receive? Why?
8. Type your name under your signature once again, even if it is in your letterhead, giving your title. Remember to keep form in your business letter. Maintaining formality indicates responsibility and reliability and your willingness to play under the general rules and conduct of business.
9. If you have an enclosure, always say so both as a reminder and in the event the cover letter becomes separated from the enclosure.

There is rarely a reason to write more than one page. Make your point and get out. A letter that wanders or tries to overpower usually fails. A succinct letter that is meaningful impresses. A letter is a way for you to show exactly who you are. Your prospect will be able to tell a lot about your work just from your letter.

Always try to send something with your letter to prove you can be of benefit to your prospect. Some enclosures that would be effective would be:

- A second current or recent news blurb or article on your work (one with a photo is even more effective)
- A circular on a particular project you have done
- A letter from a satisfied client
- A third-person mention in a letter written by a former or current client to another business person

- A newsletter mentioning your name and/or project(s)
- Anything that has appeared anywhere that casts you in a positive light

TELEVISION BROADCAST ADVERTISING

The production of television advertising is another way of achieving wide visibility for your work, as well as excellent remuneration. While the creative end of this field is dominated by the advertising agencies, few have production facilities and, thus, look to independents to make their spots. There is a lot of money to be made here, but there is also a lot of understandable competition. The broadcast advertising scene is idiosyncratic and not always ethical, so independents interested in this genre had better have their eyes wide open from the very beginning. Since the actual process of landing job contracts in the field follows the same basic pattern described throughout this book, the concentration here is on some of the peculiarities and moral issues confronted by the newcomer.

The Creative Process: What Happens Before You Are Hired

With very few exceptions, a filmmaker producing spots for advertising agencies should give up any thoughts of creative control because that area is the sole prerogative of the agency producer who hires you. You will not always like the shooting script or marketing concept you are being asked to film; just remember that your only job is to shoot the spot as directed. You should, however, have some idea of how the creative decisions are made.

An advertising spot is born when a client—usually an advertising manager of a corporation—decides with the agency account executive that television exposure or a new television campaign will boost sales. The account people return to the agency and call a production meeting to be attended by the copywriter, art director, creative director, agency producer, and media planner. If the client has suggested a marketing concept, the creative team prepares a script and artwork appropriate to that concept. Otherwise, the creative team meets separately to devise a concept for presentation to the client and the production team.

Quite often in the advertising industry, the concept is an art director's whim influenced only by some current trend. Pursuing trends for their own sake is a disservice to both the client and the producer. Generally, whenever one agency copies a somewhat unusual concept pioneered by a competitor, several other agencies are pursuing the same idea. Not only is such imitation costly, but it also proves ineffective as the airwaves are soon blanketed with a flurry of ads, all of which look the same.

The best concepts for television spots are simple and pure. They should be unencumbered by superfluous script or production techniques and free of any influences other than the desire to sell the advertiser's goods or services. Analyze some memorable spots you can recall. In most cases, you will discover a simple underlying concept.

Once the concept is approved, a production budget is devised. At this point, the agency producer will be looking for an independent production house to film the spot. Sometimes the producer will ask for bids on a project, whereas in other instances he will select a filmmaker known to have expertise in a special field, such as cel or computer animation. Your business skills and past experience may help you land a contract at this stage, but you must also be aware of your competition and certain agency prejudices.

Realities of the Advertising World

The newcomer trying to break into the advertising game may be surprised to discover that not all hiring is done on the basis of creative merit or low bids. Although they are not universal, the problems in the following discussion are widespread and can be discouraging to the independent who is not prepared for them.

Perhaps the most prevalent prejudice controlling hiring—and in fact all creative aspects in broadcast advertising—is what I call "the press of vanity." Success depends on who you know. More first jobs are obtained through connections than by virtue of talent. Once you have worked in the field, future jobs depend on your most recent jobs and how they mesh with industry trends. Clios won for spots made five years ago do not necessarily guarantee a job today. You must be current and hot.

The second most-frequent obstacle encountered by independents is what I refer to as "the hardware game." Consider the careers of these two independent filmmakers: Richard is an award-winning producer. He is a creative man who keeps overhead low by renting rather than buying state-of-the-art equipment and by using top craftspeople for each spot he produces. His competitor, Paul, has two rooms full of hardware that amount to only the basics, but it is all name equipment. He also has an attractive receptionist, a coffee pot, and a soft couch. He does all the work himself, and his spots are generally second-rate. Yet it is Paul who grabs the larger number of accounts because clients are terribly impressed by his hardware. You must not become too discouraged by this scenario; Richard does get some contracts. Although he may miss out on some of the monetary rewards of the field, he maintains the integrity of his craft.

Moral Issues

The novice independent working in broadcast advertising must also be prepared to fend off assaults on his personal integrity. In many large cities, the practice of giving kickbacks is prevalent, yet it is seldom discussed. It is so pervasive that few employees in the industry are left untouched and many become so sick of the practice they lose interest in their work and drop out. Unfortunately, the inexperienced can become involved almost without knowing it.

In a typical arrangement, a supplier—in this case, an independent filmmaker—tells the agency producer that the job will cost $5000. The producer then arranges for a purchase order in the amount of $5500. The extra $500 is obviously to be returned to the producer under the table. The filmmaker has participated in a crime; the producer is a thief. Those who fail to play this game are usually denied future contracts.

If you are just starting out, beware of illegal or unethical practices. At the end of your career, you want to have your self-esteem intact. Do not despair of making a living in broadcast advertising while maintaining your artistic and personal integrity. There is a place for you among others who share your values and appreciate your creativity. You just have to be willing to work a bit harder to find them.

9

THE LITTLE MAGIC BOXES

Not too long ago, I struck up a good conversation with a video distributor at one of the better conventions. She was a woman in her mid-40s, somewhat worn from the long day, but one who consciously revived herself as she realized she was starting to lean against the wall.

"How did you become a distributor?" I ventured. She had done it as a defense against the poor handling of her products by another distributor. She was not bitter about it, but emphasized that the distributor was "incompetent."

"In my experience," I said, "the distributor appears incompetent, but I surmised that it was just the large list they had." Book publishers sometimes operate the same way. It is a system and each product is treated at face value.

"Well, at least now, someone is selling my own products and not letting it languish in a warehouse somewhere," continued the distributor.

Which lead me to ask, "How are you handling the other filmmaker's products?"

"With as much professional effort as possible," she declared.

"Let me have your card," I said as I left.

But I already knew that the true answer to finding a good market even with a lax distributor was not in doing it yourself, whenever possible. I did not expect that this good woman would handle my products as if they were her own babies. No one I have ever met really does.

I knew that the best strategy was to promote my own products in as many ways as possible to complement and enhance the distributor's efforts, no matter how feeble that distributor appeared, once a distributor has been assigned.

DISTRIBUTION: THE HORSE BEFORE THE CART

There are many apparent paradoxes in film and video, as there are in much of business. The first paradox is that distribution must be thought of before anything else, even the conceptual phase. This same idea is also seen in the acquisition of a computer. No one purchases a computer before first considering the software application.

In the case of film or video, you must first consider the market and who will take it to that market for you. If you fail to do this, you take the risk of wasting time and money producing a product that no one will ever see.

It follows then that distribution should be discussed before addressing the type of videos you can produce to sell in the marketplace.

Because the catalog is the biggest avenue for an independent producing the type of videos I will suggest, catalog marketing will be discussed at some length.

BIRTH, GROWTH, AND SHAKEOUT

The video business was slow to start. Sony's U-Matic three-quarter-inch format cassette was used in business and industry for many years before a commercial half-inch home format was introduced to the marketplace.

If you recall the story, following Sony's development of the first half-inch (Beta) format, porn came as the first wave of popular program source and held as the dominant form for roughly three years before Hollywood movies began to emerge and eventually overtake porn in great volume. (Please follow this brief history because soon you will see where you, the independent producer, comes in.) But it was porn and America's strange hang-ups with sex as something unnatural and "evil" that created the first impetus for the amazing home VCR machine population enjoyed today.

The timing was excellent for the baby boomers, who were approaching middle age in the 1970s. From the sales of all those little magic Sony machines came the shakedown to VHS and other manufacturers. VHS was originally inferior but could record for a longer time. For that single reason, Beta has just about disappeared, at this writing, except for industrial/educational use. Chances are, it will fade altogether as 8mm and/or High 8mm continue to advance; and 8mm will *not advance* if newer, better formats, like the anticipated recordable disc, continue to come out.

But getting back to program material, for a solid five or six years, "movies" took over. Then, in the early 1990s, movies softened and a large volume of program buyers (renters included) began to turn to alternative video: sports, travelogues, gardening, hobbies and crafts, games of chance, fitness, almost anything. That is where things are at this writing and this is where you come in.

SELLING VIDEOS

The most successful of the sales efforts were by those who somehow managed to create their own mailing lists, to set up catalogs, and to sell directly to the consumer. An example is the *Reader's Digest* Association, Inc., which had a formidable list to begin with in its vast subscriber list. Even they had trouble in the beginning, aiming a little too high with cultural videos. They soon discovered that their readers were not so much into culture as they were into travelogues, Disney films, and odds and ends of crafts, gardening, and the like. The *Digest* people first isolated their reader's video tastes and then targeted them.

Another company, Kultur International, successfully took over the cultural market with operas, concerts, and, in the case of my own video title, *The Rime of the Ancient Mariner* (which was under negotiation with Dubs when Dennis Hedlund, president of Kultur acquired and absorbed this small former distributor) has since done well.

The Rime of the Ancient Mariner is an award-winning visualized reading, by the late Sir Michael Redgrave, of the great epic poem. It initially appears difficult to categorize, so retailers shy away from it. This is an *extremely* important point. A couple of years ago, I walked into the central Barnes and Noble, a fabulous book store in New York City, and asked the video department manager why they did not carry my video, which I described, and was told that it did not categorize. This is not true, and the truth is that they did not have a category set up in which the *Mariner* could fit (although it should easily be identified as a cultural, literary, or humanities video).

The point here is to remember to make it easy on yourself and become acquainted with categories open at both small and large video stores that sell alternative videos that you can produce as an independent.

If you want to become a distributor or want to understand a distributor of alternative video (the ones you can produce for income), the key is in creating your own list for direct sales or to locate a trusted distributor *before* you begin to plan, conceptualize, and produce your video.

LOCATING AN EXISTING DISTRIBUTOR

The alternative is to discover what is being sold and by whom. If you look at Kathryn Bowser's well-researched book *AIVF Guide to Film and Video Distributors* (a copublication of the Foundation for Independent Video and Film and the Association of Independent Video and Filmmakers, located at 625 Broadway, New York, NY 10012), you immediately get the thought, "Wow! Wow! AM *I* GOING TO BE RICH MAKING AND SELLING VIDEOS." Right? Wrong!

Although Kathryn did her job well and although I highly recommend the book, there are a few problems. First, many of the distributors in the book will quickly fold or be absorbed by other companies (as consolidation continues). Second, the companies appear to be as unprofessional as most of the rest of the entertainment industry. Inquiring as a producer, I made a mailing to 28 distributors chosen at random and received two responses, both negative (form letters, yet). If the nonresponding companies actually received my letters and simply did not have the courtesy to respond, they are actually saying: "We don't care about who you are, how far you will eventually go, or whom you will negatively influence on our behalf. We are obviously not intelligent, nor well managed. Therefore, we do not really care whether we survive or not."

Indeed, a friend and fellow producer has a video in distribution that is actually doing fairly well, a scholarly video but one about interesting earth phenomena. She rarely gets her telephone calls to the distributor returned. Now why is this? Since I have been in business over 25 years, I cannot accept the excuse that the individual is "swamped." It is, pure and simple, poor management. It is "negative public relations" and, even worse, negative business relations with associates upon whom they depend for their very existence.

All of this is not meant to deter you. After all, this is simply the experiences of two people, a far cry from a scientific study. But be well prepared. Be ready to experience underdeveloped personalities on your way, people who are "lost in space," disguised and posing as business people. Be ready, also, to experience dead telephones and letters marked "return to sender." Although Kathryn Bowser's AIVF reference book is the best I have come across, there is also the annual publication, Quigley's *International TV and Video Almanac.* This book is fairly expensive, but you should be able to locate a recent copy at your local library and photocopy the pages you need from it. (If it is unavailable because your local library is suffering from budget cutbacks, write to your state and federal Congressional Representatives and tell them that the library and the book is important to you. Too many people who use the library all the time, owing much to it, ignore it when they see it in trouble, unfortunately.)

BUT ARE THE LITTLE BOXES STILL MAGIC?

You bet they are. If you are really interested in becoming a video producer, the field is truly wide open. In my case, video is a small part of my work, but I accidentally (remember, when I began it was called "Dubs") located Kultur International, a company that is doing a fairly good job (although I am never satisfied, of course) selling my video. It went worldwide in 1991 after selling successfully in North America for seven years.

If you locate a distributor, they will take roughly 80 cents of each dollar that comes in. Therefore, if your video sells for $19.95, you will get $3.99 for each unit sold. That is the reality. In some cases, distributors will try to get more at first. Try not to allow this to happen. If you do, you will regret it.

Videos are created for many reasons beyond what you see for sale in even the biggest catalogs. What is happening with catalogs? Just like cable and independent TV, they are specializing. Soon, you will see video catalogs that are specialized for one type of video alone—health, fitness, nutrition, that sort of thing. Consider specializing. The best reason to do this is to establish yourself as a name in the field and to build reputation. This serves to shorten time spent trying to locate business.

Eventually, you will see video catalogs *on video* and on cable TV. It is now becoming far too expensive for many companies to put out catalogs without having to charge a price for them, which is why you now see the new business of catalog merchandisers, such as *Catalog of Catalogs*. This will eventually force companies to cablecast catalogs.

- Consider offering your services to corporations that publish catalogs and who might want to put their catalog on disc or videocassette.
- Consider producing a specialized line of videos, for example, indoor gardening of vegetables, types of fishing, or wood carving (a full list follows), and going directly to an existing catalog after you have half a dozen or more units to sell.
- Consider creating your own catalog or newsletter specializing in something, for example, foods that heal, herbs for various purposes. The newsletter would serve as a launching platform for an endless series of videos on the same subject. Newsletters do not have to be elaborate or expensive. A friend of mine, Judy Johnson, one of the foremost paper-doll artists in the United States (perhaps the world) writes and distributes a newsletter on the subject. Almost any newsletter can generate video ideas and provide the seed for a market list.

The point is that ideas for videos are endless. Here are just a few selected because, though they might exist, I have not yet seen them as videos:

All about bicycling (care, maintenance, touring)
Sewing for both men and women (reviewing the essentials of a sewing machine and how it operates; demonstrating the use of various simple or complex sewing patterns that can be taken off the video on hard copy at this writing, the PC, laser printer, and video, are merging; showing the economies of sewing your own clothes and home accessories)
Setting up a home workshop
Leather working
Candle making
Papier-maché
Weaving
Stained glass
Pottery
Quilting
Modeling, mold making, casting
Sculpture

Drawing
Painting (oil or watercolor)
Printmaking
Printing on fabric (t-shirts, etc.)
Wood sculpture
Metalworking
Drying and preserving flowers
Canning (fruits and vegetables)
Jewelry
Enameling
Picture framing
Breads and baking
Wine making
Candy making
Restoring furniture
Bookbinding
Touch typing
Garage sales
Cloth dolls and Teddy bears
Money-making ideas
Kites and kite making
Landscaping
Meats and meat cuts

There could be an entire *series* of videos entitled:

Household Hints (handy tips)
Home Legal Knowledge
Home Medical Adviser
Household Finances

The ideas, resources, for videos can be found almost anywhere. The best approach would be this:

1. Determine how distribution will be done.
2. Establish budget(s) (for both production and marketing).
3. Raise capital (either through establishing a limited partnership or finding one or more investors; at least one or two books are available on film financing; see *index*).
4. Set production schedule.

5. Complete production.
6. Begin promotion or distribution.
7. Maintain files on profit/loss business.

PREMIUM VIDEOS

There is an entirely new field of production that has opened up called *premium videos*. What is a premium video? It is a video created to support or enhance sales by a corporation, company, or institution. Upjohn Pharmaceutical, in marketing their "hair growth" product Rogaine developed videos such as "Taking Care of Your Hair Loss" and "Patient Informational Tape," the former to help create the market, the latter to help maintain it. To build a larger subscriber list, *Money Magazine* (Time Inc.) developed "Making Your Money Count" as a premium offered to new subscribers. Similar videos can be made up by almost any company selling a product. Here are a few seed ideas:

- "BMW, A Heritage of Fine Engineering," given to qualified prospects who schedule test drives.
- "Sears: Our History and Your Advantages," given to credit-card holders who spend a certain amount within a fiscal year.
- "Amazing Grace," the story of one family's good experiences under the protection of the Travelers Insurance Company's "Red Umbrella."

If there is a manufacturing or other large company (such as insurance) in your town or in a nearby city, develop several ideas and ask to see the public relations manager or, if warranted, the marketing director or sales promotion manager, depending upon your video idea. Be sure it is a good one. You can never tell how far a good idea will take you.

But remember, this is a "premium video" idea under discussion, not corporate communications. This subject will be covered in Chapter 11.

As I suggested earlier, the format for videos has already changed twice so that today there are three different tape formats and three digital disc formats. Eventually, more formats will emerge. It does not matter what the format is to the filmmaker, for the content can be transferred to any format via analog or digital processes.

To repeat, your chances as a video producer will be greater if you specialize in one area and in a particular line of development. As the world shrinks rapidly, through interpenetrating communications technologies, specialization and fragmentation are developing in parallel.

10

FEATURE FILMS AND TELEVISION

The actor was tall and powerfully built for a man in his late 60s. Muscular and bearded, he had begun in the late 1940s and. for a brief period, was considered a major star. For all the decades, his was a love-hate relationship with Hollywood.

"Hollywood is a materialistic jungle where the strong of hubris, not the strong of heart or spirit," he emphasized with clenched fist, "prey upon the weak-willed. You cannot trust a soul, not your best friends, for they more than likely have a profit motive behind their friendship. They seek only connections. Your agent your trust least because even as he hawks your flesh, he can turn and sell you out."

"The studios are tiny fiefdoms run by malevolent little Napoleons who only understand the sound of an adding machine. If you print out in black and net out, they speak your language. If not, they pass you by like horse manure on the grand highway." The actor droned on and on until he ran out of energy and his voice trailed off.

He poured me a cup of hot coffee. Sitting back, a question occurred to me, so I asked. "Suppose you got a call from Hollywood right now." Without blinking, he answered. "I'd be on the next plane as soon as I could pack."

Sooner or later—and probably sooner—most filmmakers dream of going Hollywood: producing a film that will enjoy wide distribution and critical acclaim. You may be making a handsome living churning out local television spots, training films, or public service shorts, but deep in your soul you dream of larger audiences and princely earnings. You want to make it in feature films or television.

Can an independent really succeed in these fields? The answer is a qualified yes. Independents venturing into television or feature film production must have not only a firm command of all the basics, but also the fortitude to withstand all the obstacles the system puts in their way. The greater riches to be earned are matched by the greater number of risks to be taken and compromises to be made on the road to success. This chapter will provide some practical methods for proceeding in these larger enterprises and will suggest ways to avoid the pitfalls inherent in such ventures.

THE EXPLOITATION FILM

Hollywood today is everywhere. If you want to make a picture and can raise local capital to finance your project, then your hometown can become your "Hollywood." George Romero made his classic *Night of the Living Dead* in Pittsburgh with an estimated budget of $100,000 and a local cast that was paid only after the film made a profit. To date, it has grossed $55 million.

So you can make your feature film anywhere. But what kind of film should you make? Although there are many genres from which to choose, the single formula that has proved repeatedly successful is the exploitation film. An exploitation film ("explo," for short) is a low-budget feature produced for a very specific market. Gangster films, car-chase films, horror films, and of course, pornographic films all fall into this classification. Such films exploit current social mores.

They try to ride the crest of a wave. Each film is produced with a target audience in mind. *Super Fly* and the various *Shaft* films that followed on the heels of the black-power movement are examples of the explo at its worst/best.

There are various stages of production encountered during the making of an exploitation film and some problems that can be anticipated.

The Script

Obtain a sound exploitation property. When you select a script for your film, look for a screenplay with a well-defined, suspenseful plot and some subplot structure. The story must not only keep moving, but also allow for some character development and corresponding audience identification with the characters. A human story is best: man versus man; man versus nature; man versus beast; man versus self. A film exploring the conflict between man and society usually requires more depth of meaning than an explo can handle, so avoid such themes. Remember: The exploitation film is visceral, not cerebral.

How do you find a great property? You can check with agents listed by the Writers Guild of America to find a suitable script, or you can advertise in a writer's magazine or film trade publication. The first course can cost you dearly, but the second might result in a flood of awful scripts. Protect yourself from the latter by asking initially for just a synopsis; if you are intrigued you can always request a treatment or shooting script. If you are interested in developing your own idea, consult the Writers Guild for a list of experienced scriptwriters, or talk to someone who teaches screenwriting about writing a screenplay for you.

Do not trust only your own instincts when selecting a script. After narrowing the choice to two or three, have the scripts read by several people who know how to evaluate a good story. Do not move toward production until you have found a script that excites both you and a handful of readers.

Financing

Budget your film and raise the money. After finding the right script, prepare the most skeletal budget possible, then 25 percent for advertising, promotion, and exhibition. Prepare a written presentation for your potential investors, including a story synopsis, the script, a comprehensive budget, and a projection of estimated income from various markets: first-run theaters, drive-ins, network and independent television, cable, video, and overseas sales. The presentation should also include the limited partnership agreement under which you and your investors will operate. You will need the guidance of an attorney well versed in investment arrangements and partnership contracts. Adhere strictly to your attorney's advice.

Locating investors is not easy. Your best bets are venture capital firms recommended by someone in the financial community. Explain your needs to the broker and follow through on all leads. Young, vigorous investment consultants often have among their clients several investors looking for unconventional projects in which to put their money. Professionals and businesspeople with solid portfolios frequently set aside a bit of gambling money, and a film that can provide them with a year's conversation may be a very attractive investment. Your partners are out there—you just have to dig a bit. Also keep in mind that when you are pitching to interested investors, your track record in any aspect of film production—television spots, industrials, or educational films—should be used to boost your prospects' confidence in you and the viability of the project.

Confront the Inevitable

Earmark 25 percent of the money you raise to use for advertising and exhibition. Forget about existing distributors right from the start. Sure, you can screen your epic for them, but do not hold your breath until they sign you up.

Distribution

Your advertising spots will be the key to marketing the film if you self-distribute, so take care that they are produced well. Make the spots yourself if you have background in the field; otherwise, get help from an experienced spot producer/director.

Book your spots to run during non-prime-time movies. Air time is usually cheaper then, and, besides, you want to grab a movie-watching audience. Whenever practical, sell your film during the broadcast of films with an appeal similar to yours.

To start, you will want to book your film for a one-week run. Shop around for a theater you can afford in a desirable location, if possible, and rent it. Then air your television spots for two weeks: one week before and one week during the run. This formula sounds rather simple, but it works. Do not, however, depend solely on your television spots. Send out press passes for everyone at the local newspaper and television and radio stations. Also provide passes for anyone who might be able to help you move the film, such as other theater owners.

Finally, take the time to compile a press book containing any reviews or feature articles about your opus. This is advance preparation in the event you decide to offer the film to a distributor. In that case, you may also want to show a distributor your account books, especially if the film has made some money. Success at the box office is the primary concern of distributors.

If you are truly interested in feature films, take time to read Gregory Goodell's *Independent Feature Film Production* (see Bibliography), which provides exhaustive information on the business. I recommend especially the section on distribution and marketing, which describes the realistic, if somewhat bleak, conditions confronted by independents in these areas. If you remain undiscouraged after reading, you are indeed ready to enter the exploitation-film world.

Above all, remember that you are making a commercial film for the sole purpose of making money. The more you make, the better your position. Forget about Hollywood and Oscars, rave reviews from Janet Maslin, and invitations to codirect Martin Scorsese's next film. Forget about messages, social comments, taste, and discernment. They will have to be sacrificed. The only thing that counts in the explo game is money, money, and more money.

I call the elements that go into the making of any film venture the four Ps: property, planning, production, and promotion. Those who are optimists can also add a fifth P: prosperity.

HOLLYWOOD

The big studios no longer exist. What does exist is a "handful" of "big" distributors housing musical chairs. By this I mean that each time you call on one of these companies, you are confronted with an almost new staff of executives, or "suits" as most filmmakers call them.

So what does this mean? "Hollywood" (meaning "feature films," made anywhere but distributed by the "Big Handful": Columbia,

Paramount, Warner Brothers, Universal, Disney, and Fox) is a "small town." This means the *industry* is tiny. Everyone knows "anyone." Imagine that you live in Portage, Wisconsin, a town with a population of 7,896 (give or take a few), a town that is actually larger than Hollywood. If you are anyone of prominence in Portage, you are known.

To be anyone at all in Hollywood, you must be known by *the powers that be*. Otherwise, you are a "nobody," a "non-entity," you do not exist. Being known is the *only* way to break through to a movie contract with one of the Big Handful.

Becoming Known

So, how do you "get known"?—exactly the same way anyone gets known in Portage. First, you become politicized. You "get press" or you make connections, beginning with going to the politically correct schools. These are, primarily, UCLA and USC. On a secondary level are NYU and Columbia, but this second level is not a strong one, although gaining.

Second, and of less importance than "who you know," is the slow process of establishing yourself with a distinguished record of production. You do this by persistently looking for work, the best, in the area. Nearly all craftspeople work other jobs, mostly at anything having to do with film or video, but belong to unions; and when there is a shoot, they are called, supposedly by a pecking order, but actually through political connections.

Finally, you are more than welcome to look up the various production executive's names and phone numbers (be sure it is a current reference) in annual sources such as *The Hollywood Reporter Blu Book Directory*, which you can order for about $50 from Blu Book Orders, 6715 Sunset Blvd, Hollywood, CA 90028, or via telephone with a major credit card, (213) 464-7411, Ext. 250. Then, there is the old and reliable Quigley's *International Motion Picture Almanac*, Quigley Publishing Company, Inc., 159 West 53rd St, New York, NY 10019, (212) 247-3100.

A tip! Do not waste your time trying to connect with a "development" executive. Through the years, I have attempted to find out who these people are and why they exist. Many colleagues have "told me," but the only authority "development" people seem to have is to write form rejection letters.

Being a Screenwriter

If you are a hopeful screenwriter, this is the formula to follow:

1. *Study.*
   - Read as many scenarios as possible, of "successful" classic dramas and/or comedies or whatever genre you hope to target for yourself.
   - Then read as many different books as you can on the art and technique of the scenario format.
   - Read, also, Lajos Egri's *The Art of Dramatic Writing* and *The Art of Creative Writing.*
   - Then go back and read some current scenarios. Many are available today. Various sources can be located from trade periodical ads.
2. *Work.*
   - Write out ten different concepts following what you now know about the premise and concept.
   - Outline your stories in roughly 20 double-spaced, typewritten pages, using the technique you now understand.
   - When you have ten polished outlines, have them checked by at least two professional editors for plotting, characterization, and other strengths.
   - Make any changes.
   - When you have good outlines, write ten synopses from the ten outlines. A synopsis can be as short as one paragraph but not longer than one page.
   - Write your ten scenarios.
   - When you have ten first drafts that fill you with pride, you are ready to sell.
3. *Sell.*
   - Obtain the Writers Guild of America list of agents (West, not East, there are no significant feature-scenario agents in New York City).
   - Call at least a dozen agents, and see if they will read your five best scenarios.
   - Submit them to those who clear you to send them, placing "PER YOUR REQUEST" on the outside envelope. If you want them returned, include an S.A.S.E.

This is the professional, best-way method, no matter what your background as a writer.

SUCCESS IN HOLLYWOOD

The movie business is a tough business to crack, and there are more broken hearts, by the ton, than success stories in little Hollywood. In a place where you would expect top business practices, engineering for fine quality, and professional statesmanship from script to screen, you find helter-skelter. You would expect it because of the incredible amount of money involved. Features are now averaging from $30,000,000 to $50,000,000 for no real reason at all.

A friend of mine, Michael Twaine, appeared in a film, *Cheap Shots* (cheap, indeed), which was released on a prime New York screen (where it was observed by the distributor, Helmdale, a minor player). It did not do well, nor did it garner good reviews, so it did not get a broader pattern release of some kind. The fact is that the film cost well under $1,000,000 to make, perhaps half a million or a little more, and it is being released. There are no major names in the film.

But get this: There were no *major* names in *Rocky, Fast Times at Ridgemont High, American Graffiti, Star Wars, E.T., Who Framed Roger Rabbit? Crocodile Dundee, A Nightmare on Elm Street, Dirty Dancing, Robocop, Alien, Ferris Bueler's Day Off, The Karate Kid, Police Academy,* and many more I can name. They all made over $100,000,000 combined theatrical and ancillary sales. Are you shocked about this, especially *Rocky*? Well, Sylvester Stallone was completely unknown when that film was first released.

The old notion that "big names make big bucks" has been proven wrong repeatedly through the decades; but like a Timex watch, it "takes a lickin' and keeps on tickin.'" Perhaps agents keep this nonsense alive.

The key to success in Hollywood *has* been proved out, but no one pays much attention. It is *the concept* followed with a good script fleshing out the concept. How to recognize the concept is the problem. Most distributors appoint illiterate or inexperienced people to analyze scripts. Also, the few fine analysts around might approve the script, but, by the time it gets made, many cooks have spoiled the broth. No one can keep their hands off a good script, mostly out of fear and ego, which might not surprise anyone.

In the days of the old studio system, fine, time-proven, recognized contract writers churned out those wonderful films you can see today on TNT, AMC, and other cable stations that show older films they call "classics." Actually, few are true classics; but even the cheaply made ones, like *Detour* with completely unknown Tom Neal and Ann Savage and probably made for under $25,000, still look wonderful today. Why? The script is fine—dated now, but still fine—and solid.

Whether you are targeting a craft job, screenwriting, directing, producing, or any combination of these, your only "in," with the possible exception of screenwriting, will be political. Do not expect that having a great resume, but knowing no one, will get you a job. One out of some 10,000,000 applicants might luck out that way.

HOW TO FAIL, SIMPLY AND BEAUTIFULLY, AND GIVE THE INDUSTRY A BLACK EYE WHILE DOING SO

If failure is your cup of tea, and many observers of human behavior say it is for many people, follow this advice:

1. Dream your way into a feature with the help of a rich relative or a greedy and not terribly bright relative or friend who has financing access. Do not concern yourself with "paying your dues" and learning any part of your craft well. You might find yourself in creative charge of a picture and at the center of a shoot, on location, with 132 people waiting for your decision and have a panic attack because you really do not have the skill or knowledge to pull it off. Then, by all means, take some drugs to get you through, to give you the unfounded courage you need.

   (Am I making this up? No. I have experienced exactly that scenario.)
2. Similarly, when you determine you have what it takes to mount a feature after winning an award or two for a couple of ten-minute 16mm films or school videos, write your own script, insisting that you know more than a seasoned veteran and member of WGA. Do not seek a good script; do not seek the help of any professionals.

Go out and make your own bomb, and waste the hard-earned money and trust of your investors. Do not care what kind of bad taste you leave in the mouths of these simple-headed investors. After all, they were dumb enough to put money on you without checking your background, right?

Today, more than ever in my experience, young people are prone to be less interested in the art of the craft and the doing and more interested in the winning. This does not happen in sports because no one goes out there and wins without hard work. It does happen in the high-stakes feature-film industry. In some case, by some incredible luck, an incompetent will score on a first outing and then finds financing for a whole series of films, only to go out and lose big with each and every one. Yes, this is a strange business. Strange indeed.

The Independent's Role

The scope of cable television is so broad that the independent must actually choose a field in which to concentrate. Once you have chosen a category and developed a possible film subject, contact the acquisition director of the appropriate cable service or services and find out how the service acquires programming. Stress the financial details. You will probably discover that at present few services have money to sink into production. Programming is generally acquired from the major studios. Therefore, if you are determined to produce for cable television, you must form your own production company and seek funding from investment sources. You can find a market for your work if the quality is good and you have done your homework.

Another area open to independents requires much less risk-taking. Almost all cable systems are in the market for shorts or intermission fillers. Such films can run anywhere from one to thirty minutes, with ten minutes the preferred length. Favorite subjects for shorts include comedy, animation, Hollywood star interviews, sports, and seasonal subjects. When filming a short, be sure to capture the viewer's attention within the first minute. Also keep pacing and production values at broadcast-television level. Forget about experimental films unless they are unusually clever and capture the public imagination. The acquisition director can provide valuable advise as to the most salable topics.

The exploitation film, and broadcast and cable television all offer the independent the possibility of mass audiences and occasional acclaim. However, these fields are complicated by a seemingly endless array of problems. As a filmmaker, you must decide whether or not the obstacles one must overcome are worth the rewards. A good living can be made by those who enter these fields prepared to take the risks and to work hard.

11

THE CORPORATIONS, LAND OF REAL OPPORTUNITY

He wore a three-piece suit with rounded edges that appeared as if it had been pressed while he wore it. Not that it was crumpled, but Jay Gatsby would not have been caught dead in it. He was a tall man, yet his self-effacing ways made him seem of average height. Often, he would sit alone in his car and smoke a long Havana cigar via Canada and snicker to himself. My fellow creatives, other writers and directors, would nod toward him and draw circles around their temples with an index finger and mumble, "That's what eventually happens to all account men."

But Ted Westermann, a wonderful gentleman who has long since faded from the scene, was not someone you could just shrug off. He was very successful, and, after some months of observation, I approached him in an effort to find out why.

During a lunch break on a set, I saw him laughing to himself again and decided to intrude upon his little magic world. "So what's so funny?" I asked, not waiting to be invited to sit down. "Oh, nothing . . . nothing at all," he said, still breaking up. I envied him his secret joke. We talked one hour into many hours during the passing months and became friends. I found that Ted, far from being a candidate for a fruitcake factory, was quite brilliant.

Ted was the man who called on General Motors. He did it with great skill. He constantly brought in business to our large studio, which needed a lot of business to keep going. The thought of working a large company like General Motors intimidated me. I asked Ted how he did it. "No problem," he said, "they're just people like you and me." Not satisfied, I asked, "Well, how do you keep clients eating out of your hand like you do?" Again, there was the Ted Westermann smile of confidence. "Ya gotta be funny. People have all the grief they need already. They need someone to pick them up and dust 'em off. All's you gotta do," he said, patting the ashes off the large Havana with the flamboyance of a conductor before the Boston Symphony, "is have a good sense of humor." My God, I thought, how can a creative guy like me who is always playing a Greek tragedy have a sense of humor? "If you want to be popular anywhere . . . ya gotta have a good sense of humor," he asserted.

Ted was right. At times in my career, I had to rediscover this great secret the hard way, but, for the most part, what he told me stuck through the years. As business flowed under the bridge, it became easier when things got tight to maintain a sense of humor. Soon I realized that no matter how big a corporation was, it was still composed of people, just as Ted said. This was the most important thing to remember when considering large accounts without the help of a wonderful Ted Westermann. He is sorely missed.

Throughout this book have been explored the multiple and varied business opportunities available to the filmmaker willing to use some imagination and hustle. The canvass of potential clients has included community institutions, small businesses, advertising agencies, and cable television. However, the single greatest source of jobs for independents, large corporations, has been mentioned only fleetingly. Almost every type of visual communication described can be used by some division of a large corporation. The following discussion of the filmmaker's role in big business should indicate the variety of jobs that can be gleaned from this source—if you know how to find them.

HOW TO ATTACK THE BIG CORPORATION

It was suggested at the end of Chapter 5 that an organizational chart is a useful tool to have when you are preparing to solicit business from any large corporation. Now take a closer look at the typical corporate structure in order to consider ways in which you as a filmmaker can sell your services at several levels and to several divisions within the same company.

In preparing to approach a corporation, think of yourself as a communications consultant and act accordingly. You want not only to attract business, but also to promote your services. Consider what a corporation does. It manufactures a product or prepares a service and then markets that product or service. Although most communication will be done during the marketing phase, do not overlook the opportunities available during the production phases, perhaps in training or internal communication. Seek out the individuals in charge of each division and find out how they do their work. You may discover ways in which you as a filmmaker or audiovisual producer can help.

The chief executive officer of a corporation is usually the president and can be found at the top of every organizational chart. The filmmaker can generally bypass this office when contacting a large corporation. Concentrate instead on the officers who report directly to the president in the areas of operations or manufacturing, marketing, and finance. These operating vice presidents and their immediate assistants—managers and supervisors—are responsible for purchasing screened communication from outside vendors and are, thus, the proper target for the filmmaker's attention.

The corporation offers not only a wide range of business opportunities, but also the possibility of a continuing account. To maintain such an account, your understanding of the corporate structure must go beyond the organizational framework to allow for personnel changes. During the course of a continuing account, you may find that, due to personnel changes, you will be working with different executives than those with whom you started. As a result, you may have to modify account strategy to reflect differences in the philosophies and personalities of the new people. Avoid stepping on toes by working within established channels; always present your advice and suggestions to the appropriate person. Steer clear of corporate politics. At whatever level you have access, maintain your posture as an interested and friendly observer of the corporation.

If you are a keen observer, you will be able to locate film opportunities within the corporation beyond those associated with the marketing division. Personnel departments in large organizations are quite sophisticated and eager to use modern communication methods to handle such functions as orientation, behavior modification, safety campaigns, and the introduction or review of corporate policy. Video, film, or audiovisuals are also favored for training purposes. Keep in mind that such jobs may be commissioned either by the personnel office or the training director of a manufacturing division. Although such films will not garner the big budgets associated with marketing, they can keep your shop busy at times when you may need new business most.

THE SPONSORED FILM

The public relations office is another source of business for the independent filmmaker. The public relations executive is responsible for communicating the corporate image to employees, the business community, and the general public. Such communications take many forms, and sometimes the chosen medium is the sponsored film.

A sponsored film is a public relations film or a public service film made by a corporation or institution to promote a positive image for itself. Although the film need not focus on a company's product or service, it will usually have a message or motive beneficial to corporate aims. Such films are presented gratis to the public. Distribution can be handled by the sponsor but is generally undertaken by a professional distributor who is paid by the sponsor for each audience screening. Each audience is also surveyed in order to give the funding company data on the film's use.

The most important aspect of making a sponsored film is determining and implementing its distribution. There is little point in obtaining budget money to produce the film unless it is backed by a good distribution plan. After all, a company that sponsors a film that never enjoys widespread screenings is unlikely to approve future film efforts. If you need help in this area, consult the two companies that currently distribute sponsored films: Modern Talking Picture Service and Association Films, Inc.

Do not forget to separate the budget for distribution from the production budget. After the film is produced, the budget for production can be laid to rest, but the distribution budget may need to be carried through several fiscal years. A highly successful sponsored film may be screened for many years, thus continuing to create a positive image for the sponsor.

The sponsored film is a great opportunity for both the sponsor and the filmmaker. A company is provided with a worthwhile and beneficial project in which to invest. For the filmmaker, a well-made sponsored film can result in new sources of business as well as personal fulfillment and satisfaction. Work carefully if you get a chance to make one, and, by all means, read Walter Klein's excellent book on the subject, *The Sponsored Film* (see Bibliography).

THE MARKETING PROCESS

The most obvious place for the filmmaker or audiovisual producer to look for business opportunities in a corporation is the marketing division, yet even there hidden possibilities abound. Far from being limited to commercial spots and sales presentations, a role may exist for the visual communicator at almost every level of the marketing process.

Market Research

Market research is the first stage in the marketing process, especially when a new product is introduced. Researchers analyze surveys, company records, industry data, census figures, and other factors to determine the market potential of a product. Because corporate investment in further development and marketing of a product hinges on accurate market research, procedures and data are continuously verified, reviewed, and updated.

Although market research has not made use of visual communications thus far, the potential is certainly evident. By videotaping surveys, for example, it would be possible to analyze not only answers to questions, but also body language, facial expression, and, in fact, every visual nuance. An enterprising filmmaker, possibly teamed with a behavioral scientist, should be able to find a lucrative career in this field.

Planning

Planning, the second phase of marketing, establishes the company's mission, scope, and goals by producing a situation analysis and determining marketing objectives. A resourceful audiovisual producer can provide a visual presentation of the marketing plan to be communicated to marketing staff and corporate executives. Slide shows accompanied by a tape or oral script are preferable to film or video in this situation because they permit flexibility in the continuity of the visual presentation and readily accommodate frequent modification.

Once the market plan is completed, product planning and development (PP&D) begin. Here again, the audiovisual producer can play a role in the screening and appraisal of new product ideas or actual product prototypes. Presentations can be simplified further by using closed-circuit television where available. Whereas the final management review formerly rested upon a written planning-and-development reference guide, the videomaker or audiovisual producer can now offer a videocassette reference guide backed with figure charts.

Distribution and Pricing

In the next step of the marketing process, product distribution channels are established. Any mediamaker should be alert when distribution is mentioned. Distribution means communication, and effective communication today means some type of audiovisual. When developing a distribution network for a new product, executives consider various alternatives and then select a specific type of distribution, as well as appropriate salespeople, agents, and distributors. Once a written plan is developed, the entire concept is evaluated and, if necessary, reworked. Finally, a distribution-channel reference guide is prepared. At each of these stages, the mediamaker has an opportunity to assist with corporate communication.

Once distribution has been determined, pricing is established. This is the realm of industry specialists, and no realistic role exists for the filmmaker at this stage. After pricing, however, the company will engage in advertising, public relations, and a sales campaign. These are the natural markets for the filmmaker, whose function in each phase has been discussed extensively earlier in this book.

Campaign Appraisal

After a campaign is run, its effectiveness is measured by various means including incoming letters, inquiry reports, readership studies, awareness and/or reference studies, and other market tests. Depending on the operation, a visual presentation may be appropriate. Consider inquiry handling, for example. When an inquiry is received on a new product, it represents a prospective client; such leads must be communicated to the sales force promptly and handled by them efficiently. The knowledgeable filmmaker can sell the sales promotion manager a film, slide show, or tape that emphasizes the importance of correct inquiry handling. The filmmaker can also seek funding for an industrial educational film on inquiry handling and offer it to other corporations as a refresher course.

The legal aspects of marketing can also benefit from the input of a capable filmmaker. Marketing legislation, patent and trademark protection, distribution contracts, and consumer, fair-trade, and ecology laws must all be reviewed constantly by executives; yet many companies are unequipped to provide such information. You can help by proposing an annual slide and/or tape review, which would be researched by you or a freelance writer. Considering the increasing complexity of the marketplace and heightened consumer sensibilities, the market for such audiovisual services should continue to grow.

By now, it should be apparent that big business offers multiple opportunities for the well-prepared filmmaker or audiovisual producer who is willing to present innovative approaches to traditional problems. In this respect, the independent filmmaker relates to the large corporation in exactly the same manner as any other client. The strategy remains the same whether the end product is a television spot, sales promotion videocassette, or sponsored film. The goal is to find business and fulfill yourself as a mediamaker while serving your client's best interests.

A good living can be earned by the independent filmmaker. Financial success is currently achieved by finding and nurturing a wide variety of clients—by hustling and always producing quality work. The road for the independent may ease in the future. Videodisc and other advanced technologies may permit filmmakers not only to control the artistic content of their work, but also to allow them to distribute their films directly to the public, thus reaping the appropriate compensation. Until then, it is enough that filmmakers with talent, business sense, and fortitude can make money while pursing a craft they love.

12
WHEN CORPORATIONS FAIL

We sat in an old hotel bar somewhere in Portland, Oregon. Our flight had been canceled, and we had an additional three hours to kill—packed and ready, having checked out of our rooms. At first, I took no note of him, but as I glanced around the room, he sat there smiling at me, and his face lit the otherwise gloomy room. After a few more occasions of our eyes meeting, he waved at me and motioned for me to come over.

Skippy Tyler, as his name turned out to be, was twice my age, with what appeared to be twice my energy. He had missed the flight along with a handful of others whom we could easily pick out of the crowd.

We decided to visit and introduced ourselves, throwing in a brief on our backgrounds. "Oh, I've worked for them all, mostly companies that once were household names, companies 'You could rely on,' which are now gone forever. Packard Automobiles, then DeSoto, Nash, American Motors, and after a while I moved to a 'safer' industry, the business machines field. There, I worked for solid *companies like Marchant, L.C. Smith, and Underwood. You remember Underwood, of course, the old Standard Five? I sold thousands of them to schools."*

It was difficult to be serious, because I knew Skippy was being facetious. The first question was the most obvious, which I did not ask, but he answered, "Why did they go out of business? Hardening of the arteries," he said, smiling.

It was what I had come to understand and although it is an easy formula, that somehow communications break down, through the years, I began to see a clear life cycle in companies. The founders had the heart, spirit, drive, and ambition, and they were eventually replaced with incompetent people who slowly moved up to power by not taking risks. Simple? Most large companies today in the United States are in the white-elephant stage of growth. They are on their way out. IBM just fractured itself into pieces in order to survive—a valiant act, and one that might pay off. IBM has been fortunate to have mostly decent management all along, and it would be interesting to study the biography of this company.

Having worked for General Motors while with Jam Handy, as in most of the companies I serviced through the years as a communicator, I had developed a loyalty. In the early 1970s, I wrote to the GM chairman at that time and identified myself as a former Jam Handy staffer (which was always well received at GM and for good reason). My letter was about how terrible I felt their products had become. I warned him that the Japanese and Germans were producing cars of far better quality. His response was a superficial sales-promotion letter.

REFLECTIONS

If I were two decades earlier in my career, this chapter would have been a composite of other people's writings, complete with attending, statistically significant figures to assure you of validity. You would be given a host of differing possibilities, many of which you might never encounter throughout your entire career.

Instead, it seems more appropriate to offer firsthand information and some history on my interpersonal relationships in the business, many of which I am quite certain you will experience, if you have not by this time.

Indeed, there is a large library of pop psychology today, nesting easily upon clinical work going back, now, nearly a full century from the time psychiatrists were called "alienists." Among the books on pop psychology, there is a series of books on the creative mind and its survival in the everyday workaday world.

The reason this chapter straddles both political thought and psychology is that they are completely intertwined. It is all actually quite simple. One can distill it all down to a central issue, even down to one word—*fear*. Most of the problems you will ever face in the workday will be based upon someone's fear. If you can keep this in mind when confronted with it, and not react with your own fear, you will go a long way toward succeeding among your peers.

THE REAL WORLD VERSUS THE IDEAL

Growing up, I was fortunate, in one sense, to have a disciplinarian father. He constantly threw little bromides and analogies at me and my brother, Ray (who is a cel animator today). Father, unfortunately, told Ray and me that hard work pays dividends. Actually, "hard working" is a given for any shot at success, but it guarantees nothing. Thus, his advice was very basic, simplistic, and lacking, setting me up for a big fall. He also said that honesty and dedication are "sure-fire" ways to succeed. Well, the same goes for that. There are NO guarantees, NO work in an idealized reality.

Quite often, a person is faced with an experience that defies all logic but that can be explained in terms of simple human frailty. The following is an example:

The Life Cycle of a Company

Looking around at corporate America at the end of the twentieth century, some giant white elephants can be seen among the herd. These are beasts that have grown so fat and tired, they cannot stand out of their own way. Their fate is sealed, and only a miracle will keep them from failing, in some cases, to be quickly devoured by waiting jackals, just like nature.

My first job in films was with a white elephant, only I did not realize it when I first went happily to work there. The Jam Handy Organization was one of the greatest film companies (if not the greatest) that ever existed. It was, historically, the second business-film company in the United States, founded in 1910 Detroit just a few months after the first one, which was the Reid Ray Film Company in, of all places, St. Paul, Minnesota.

By the time I went to work at Jam Handy (JHO) in early 1965, it had offices in four cities, including New York City and Los Angeles, employing roughly 400 to 500 people. Most of the operation, however, especially production, was housed in Detroit in a series of buildings sprawling down East Grand Boulevard, running northeast from the General Motors Building. The various GM divisions were some of JHO's major accounts, including, of course, Chevrolet, which was then their largest single account.

The founder of JHO, Jamison Handy, was a true legend in his own time. He had been a sickly child, having experienced polio, and had been partially paralyzed. He prevailed, however, and took up swimming in order to regain full use of his body. His faith in this concept, as a Christian Scientist, was strong. Jam's powerful will was in his great swimming achievement, for he became an Olympic bronze medal swimmer in the 440–breast stroke at the 1904 Olympiad, just one year after the Wright Brother's famous Kitty Hawk flight. The very next year, in 1905, Jam Handy innovated the freestyle-stroke breathing technique that is used to this very day throughout the world.

Nearly 20 years later, while at a luncheon at the Detroit Athletic Club, someone ventured to suggest to Jam that "it was just too bad he was now 'over the hill' for Olympic competition." Jam went out again and won a position with the 1924 Olympic water polo team, at 38 years of age.

As the decades rolled by, he became the "Dean of American Swimmers," and among those inspired by him were Buster Crabbe and Johnny Weismuller, both of whom went on to become actors and movie Tarzans. Buster Crabbe is also remembered for his early "Flash Gordon" serials, and Johnny Weismuller for "Jungle Jim" on television in the 1950s.

Having conquered his own demons, Jamison Handy went on to instill American corporations and institutions with this missionary zeal and positive thinking. To a degree much greater than most educators in business schools can today recall, including Harvard and Stanford, Jamison Handy was a large part of the American business success story in the first half of the twentieth century.

With only his experience with the *Chicago Tribune* behind him, where he was reputed to have held as many as 72 different jobs in seven years, Jam Handy essentially invented sales promotion, merchandising, and inspirational, motivational, and instructive audiovisual aids in business. He also developed a little thing called the "filmstrip," having invented the gate for the projector. Filmstrips are still in use to this day worldwide. He focused on the audiovisual field because he was so sold on these neural pathways as the keys to transformational behavioral changes and learning. He had a saying, "Nothing goes in one eye and out the other."

Mr. Handy set in motion audiovisual training in the corporations and institutions of this era. In 1913, he produced the first "live" business theater for National Cash Register. With the advent of World War I, he began producing filmstrips, another worldwide first, for World War I arms and submarine training.

In 1929, General Motors had a division, that competed against the Ford Motors powerhouse. This division, named after the early race driver Louis Chevrolet, was very weak, but it had good products. GM was thinking of dropping the entire line. After some research, Jam found the trouble. Most of the salesmen were intimidated by Ford; some even owned Fords and "swore by them." He also found that the Chevrolet line was as good if not better than the Fords, which were ubiquitous in that era. With Jam Handy's unconquerable spirit, the troubled Chevrolet Division not only turned around but became dominant in its class, where it remained for most of the balance of the century. He did it primarily by strengthening, inspiring, and motivating the dealer division. He already knew that companies fell apart due to lack of communications. Jam Handy would laugh at today's American fear of Japanese products, which though good, are not omnipotent.

When World War II broke out, he went on to do the same for the armed forces of the United States and was recognized for his innovative work in informing, inspiring, and motivating the troops. By the end of the war, his company had produced roughly 7,000 flight-simulation and target-practice films.

It was this great effort of his that changed business for all time, for Jam Handy lead the way the world over in motivational, inspirational audiovisual aids, and group training.

I first met Jam Handy in 1966, and, at 80, he was still a strong and vigorous man. In those days, he would say he would "live to be at least 140, not quite a Methuselah, but I'll give him a run for his money."

Jam Handy did prevail until 1983, just three years short of his centenary. In the later years, somewhat remote from the firm he had founded so long ago and from the changing times, his company foundered, as I mentioned above. But shortly before he died, Jam helped Bill Sandy create Sandy Corporation. Bill was a soft-spoken writer who worked a couple of cubicles away from me in the old

1960s Jam Handy Organization editorial section. Sandy Corporation is a "high tech" version of the old JHO and doing well to this day in Troy, Michigan, carrying on the basic traditions that Jam Handy innovated back when the twentieth century was young.

For five decades, this film and audiovisual company had prevailed, becoming the largest of its kind, spinning off a handful of smaller companies. An interesting footnote, Max Fleischer, the wonderful animator who, in the early part of the twentieth century, had inspired the young Walt Disney with his *Ko-Ko the Clown, Out of the Inkwell,* and other creative brainchildren, worked at JHO during World War II, helping with the various training films at their animation unit. Max had created Betty Boop, Popeye, Superman, and Raggedy Ann and Andy animated films, among many others during his golden days.

Max joined Jam Handy right after a Paramount fiasco. This big studio's executives, at the time, were intimidated by the great Walt Disney features of the period. It should be mentioned here that Disney had released *Snow White, Pinocchio,* and *Fantasia,* by this time. Paramount decided to orphan the only two features ever produced by Fleischer, "because they were not up to Disney standards." These two fine Fleischer features were made in Miami, Florida (where they went to escape the labor union). What the Paramount executives could not see is that the Fleischer films were quite good, but done in a different style. The films were *Gulliver's Travels* and *Hoppity Goes to Town,* two favorites of mine, which I feel are better than many of the more plastic Disney films that followed.

Paramount's stupidity of the period immediately following seems also to have known no limits; perhaps, because like the rest of corporate America, Paramount esteemed short-term profits only, selling off 2,000 Fleischer shorts to television distributors for about $4.5 million. The shorts have since brought in many times that amount and still have unlimited legs, quite likely to gain in value with time.

However, Max Fleischer's work at JHO was not good. The JHO animators, an excellent unit, were chagrined to see this. There was one film, used heavily on early TV, that Fleischer did at JHO, *Rudolph the Red Nosed Reindeer,* a film dusted off for TV every Christmas. In any event, JHO and Max parted company by the early 1950s.

By the time I arrived at JHO in 1965 to work first at various departments in an unstructured management-development experiment, Jamison Handy was in his late-70s. The once brilliant studio was now suffering from the "cobbler's children" syndrome—while they produced fine communications for top Fortune 500 corporations, they themselves were terribly managed.

After stints with the animation unit, the editorial department, and a group called Field Service, I joined the production department. There, I saw incredible waste. There was, to cite one example, a glass-walled room filled with photographers "on call," having punched in their cards, sitting, playing cards all day long. With my work-ethic orientation, I applied myself so diligently that they began to heap me with work. One month, I noticed on my routing board 136 different job numbers, including 100 or so updates.

After having a filmstrip continuity shot with some frames upside down, I talked to an elderly man named Billy, who was the cameraman at the time, and noticed he was nearly blind.

At this time, I began to submit "efficiency reports" mostly offering cost-cutting ideas and suggestions. Although I was not tutored in business (my degree is in psychology) and was no industrial engineer working in time and space studies, it was all so illogical to me that I became frustrated when management did nothing to improve conditions. Executives drawing top salaries would come in, *Wall Street Journal* in hand, have morning coffee while waiting for lunch, then return from a long luncheon in midafternoon with the late-day paper in their hands. Twice a week, they would sit and review films and other products for "quality." In the meantime, "the old man," Jam Handy, was in abstentia, living in Carmel, California, some 2,500 miles away, while I ran, sweating, down the long studio hallways, carrying various production pieces and watching the place fold.

My vision was that of the entire company sinking into the sea, like the *Titanic*, taking with it, the 400+ families that relied on JHO for a living. I was called on the carpet for complaining and offered a "promotion" to that of an account executive. After two or three contacts with clients when I asked about their product satisfaction, price competitiveness, and so on, I ran back to management with even more complaints, telling them the company was over 100 percent overpriced against competition. I was told "We are Jam Handy; we cannot be overpriced at any cost," bizarre things like that. It was as if everyone was oblivious to the dangers, and I began to get migraine headaches over this, being so stressed out.

At one point, I was told that Jam Handy had been around for half a century and that I, much too green and inexperienced, did not know nor could know the entire picture (which to me was as occult as the giant bromides printed on the outside studio walls, "JHO, to inspire, to motivate").

Frustration led to my rocking the boat, real hard; and, eventually, with no response, that led to my attempt to make my fellow workers aware of our mutual danger. Everyone turned the other way, except for my immediate supervisor, Pat O'Hara, and his immediate supervisor, Lyn Wellhausen; both competent men, but helpless to do anything. After a brief ceremony, one fine day, I was thrown out of the company.

One year later, I was with Cessna Aircraft in Wichita and had gone back to Detroit to work on a sound track with one of the sound studios that used to subcontract for JHO. I decided to visit my old barber across the street from the central studio on East Grand Boulevard. After getting a haircut, as I was walking down the street, headed to my appointment with the sound studio, I heard my name being shouted in the street. It was Lyn Wellhausen. He told me, "Raul, you've probably heard, we're going out of business." "Yes," I said, "I heard." Looking into my eyes, Lyn said, "We hired a very expensive management consultant half a year ago. They've been here all this time and they have yet to turn in a single report I could not match with one from your old file, which you had turned in, unsolicited, gratis, to us, while you were employed here at Jam Handy. Well, I just wanted you to know that," offering me the only tidbit of compensation I ever received from this historic, once-wonderful company, for which I suffered greatly, losing my job in my failed attempt to save them.

The point to this entire story is that a company is only as strong as its weakest link, and there are people who cannot help but put their frail egos before the welfare of hundreds, sometimes thousands of people whose livelihood hangs on the daily decisions of a weak management.

Was this a worst-case scenario? Not by a long shot. Workers are often confounded by the sheer stupidity, avarice, and fear in fellow workers, supervisors, clients, and so forth, who find themselves in positions of management to the utter destruction of their companies.

Everyone has heard of the "whistle-blowers" who turn in incompetent or even criminal management only to find themselves quickly out on the street while the exposed managers go on to promotions and higher salaries. It has happened in my experience more than once, and it is a reality of the workday the world over.

There is nothing that I can tell you to help you face this inevitable, illogical conflict but to deal with it with courage and take a chance to do the right thing. The bottom line is that you have to face yourself at the end of each day and at the end of your days.

EXPECTING THE BIZARRE

Idealists will tell you to "expect the good," and they are correct. For, if you look for a dark cloud, the chances are that is exactly what you will experience. Positive thinking is the only real choice. Whatever bizarre encounters you have, consider them as part of the pattern, just as you will see darkness in a large painting only to contrast the highlights and make up the composition.

In film and video, clients are more impressed with hardware than talent. In actual fact, many of them cannot judge real talent, so they base their judgment on expensive hardware. The filmmakers with the best hardware usually get the job order. This weirdness is historic.

In the same way, people tend to shy away from risk, even when it exists only in their mind. One laughable example is taken from my brother, Ray, who is one of the finest animators alive in the world today, and I would not say it simply because he is my brother. He has worked on many television specials, features (with people like Dick Williams and Shamus Culhane), and dozens of commercials. One day, he was called to show his reel at an agency. In the reel was animated all manner of objects (from an automobile carburetor, for example, that was given a lifelike personality for Pennzoil, to rabbits, ducks, humans, and so on). The agency people asked, "You don't have a turkey?" Ray responded, "No, but it's easier to animate a turkey than a duck, which I have on the reel; a duck waddles." To make a long story short, they gave the job to someone else, not understanding that a good animator *always* studies the movement of his subject before going to work. An extraordinary animator can create an exaggerated characteristic motion or invent it, as in the carburetor in the Pennzoil spot.

Often, you will see people hiding their lack of ability and cowardice under bombast, lies, and insolence. Sometimes it can be quite subtle. Rarely does the offender realize it. Although this takes place everywhere, it is most poignant in filmmaking where so much depends on clarity.

13

INTO THE FRAY

There is a man I know very well. He quietly helps others without fanfare or notoriety. He considers everyone he meets an opportunity to raise his own self-esteem, and he does this merely by being considerate.

In one out of many examples, he once hired a freelance editor, and by the time the editor was half-finished with his job, he discovered the editor was going through a difficult financial period; his cash flow was nearly flat. Without any conversation, he paid the editor for the entire job in advance. The editor knew that most people in our business often pay craftsmen from one to three months after the job is delivered. He also knew this man paid everyone cash on delivery, often undergoing personal hardships to do so. The editor never forgot the man and, nearly 20 years later, became the man's supportive partner, a partner whose dedication and devotion was unshakable. From a single thoughtful act, both men and everyone they touched, were affected positively.

THE UNMITIGABLE KEY TO SUCCESS

Recall what success is: It is not how much money one accumulates in a bank account or investment portfolio; it is, beyond anything else, achievement of a goal and fulfillment amongst peers. No one believes in their right mind that the men who went to jail for insider trading, many with their booty intact, are successes. They are dismal failures before society. Wealth alone is never a true measure of success.

As in the anecdote above, a successful person is the one who has created riches of the heart, much like George Bailey in *It's a Wonderful Life*. The cynics can call it "cornball" or "naive," but this formula will prevail as long as there is a society of humans on earth. Knock on what wood remains.

THE FUNDAMENTAL SOURCE OF FAILURE IN BUSINESS

In 1964, the last company I worked for before moving on into filmmaking with Jamison Handy was as a consultant with Yale Laitin Associates. This is a company that does corporate and institutional attitude surveys.

An attitude survey is performed by a team of people that is independent of the corporation and usually hired directly by the CEO. First, the team asks random questions of various employees in a company. Then, from studying the random samplings, they design questionnaires, color-coded from one group to another, from one division or department to another. All the questionnaires vary and are designed to address particular problems that were revealed by the random samples. The questionnaires are sent out to each and every employee in the company everywhere, from the very top executives down to the lowest paid messengers and mail clerks.

After the questionnaires are returned, the answers are entered into a data base and are then collated per department, converted into meaningful statistical data, and printed on large spreadsheets that show detailed attitude profiles. This information is then used by the CEO to implement expansion, reductions, and changes to maintain and reform the company.

During my brief tenure with Dr. Laitin, I learned the answer to the single greatest mystery behind business failure—an answer so simple and so obvious that it is almost always overlooked. (But people tend to seek complicated answers to complicated questions when, always, the answer is simple.)

What is the single source of failure in business? A breakdown in communications. Time after time after time, when doing attitude surveys with companies anywhere from gigantic companies like General Motors (which had them done every few years in those days), to ivory-tower pharmaceutical companies such as Merck, to modest industrials such as Cooper Tire and Rubber in Findlay, Ohio, to small operations such as the New York State Thruway Authority, the problem was always a break in communications of some kind.

The attitude surveys were good. They provided a bridge from the entire body of employees to the executives at the very top, without political treachery or personality bias. But they could be only as strong as the weakest CEO fearful of changing the status quo.

LEADERSHIP

Are there weak CEOs? One would logically believe not. After all, these people are sometimes responsible for the livelihoods of tens of thousands, and their influence often reaches hundreds of millions of people. One would believe that someone, at some time, devised a corporate system of executive selection based on talents and effectiveness. As I discovered, much to my amazement, many CEOs arrived at their lofty positions by default. Too often they were people who remained in a company by staying out of harm's way through cowardice, guile, or both; or they moved up politically.

Hollywood is no different. The few CEOs who attain their positions through sheer performance and raw management talent more than make up for the rest, and they are inspiring to behold. But their number is small compared to the rest.

In Chapter 12, I showed how the wonderful and historic Jam Handy Organization failed through lack of communication from the top down and an idiotic refusal by management to listen to the suffering "peasants." This curious pattern precedes nearly every revolution; but unfortunately few "leaders" are truly educated, and history repeats itself again and again.

When someone puts a job on the line to try to save a company, the individual is almost always fired, like the legendary whistleblower. The company usually goes on to self-destruct with the ensuing downfall of, all too often, great and historic institutions to which many gave their blood, sweat, tears, and sometimes their lives. The net result is the displacement of hundreds, sometimes thousands of families, a phenomenon that sometimes negatively affects entire national and even international economies, culture, and history.

But why will a company self-destruct? Why does no one seem to listen to the single individual in all too many cases? Apparently, it depends upon the political positioning of the single savior. If he or she is not well placed and in a strong position, strong enough to overcome the ironically nonproductive atmosphere of fear cultivated by many company managers and top executives, the company will go down.

The sole savior, as seen in the life of Jesus Christ and in many other stories too numerous to list here, usually self sacrifices. This is a curious aspect of the human drama. Often, the triumph of the human spirit is seen, even though the individual may have been sacrificed. In the end, the drama serves spiritual growth in its telling.

The best screenplays and movies are based upon universal themes. This same drama takes place in films like Alfred Hitchcock's great 1944 picture, *Lifeboat*. Here are a group of survivors who most certainly would have died if it had not been for one among them who took the helm and, though eventually hated by most of the survivors, made it possible, by virtue of his firm and logical leadership, for most to survive the horrible catastrophe.

So why are you being told all this? Most of your work in film and video will be or is to communicate. Many of you will make a difference. Some of you will become entertainers. Even then, you will prevail primarily through your thoroughness in communicating to your crews, to management, to the talent. Communication has moved, in the last part of this century, from print to electronic and film media. You are where the action really is, in communications.

THE FUEL FOR SUCCESS

Simply put, success is achieved through inspiration and motivation. Inspiration and motivation do not happen automatically; they are *made* to happen. It is done with homework. And there is help, lots of it. The best organization for assistance in the field of motivation and inspiration that I know is Nightingale-Conant Corporation. Their toll-free catalog number is 1-800-525-9000. Review the catalog and make an investment in your future. Most of Nightingale-Conant's programs are excellent. Use them throughout your career and begin to notice the changes in your behavior as you strengthen your will and create new levels of self-confidence and concentration. Note, also the amazing, positive changes that take place on a personal level, not only at work, but at home with your family and in your community.

Managing, conditioning, strengthening, and maintaining control over the mind is not something in some far distant future. It is here and now. It will be a great force in the twenty-first century at every level of society if it is to survive.

So this is the end of Chapter 13 and the end of my visit with you.

Go out and make good films to create a good world for yourselves. You have the power to do it.

This is, more than ever before, the time of communications. The great many opportunities existing today for film and video are brand new.

As last word, trust the power of a smile. Whatever you do, never forget to smile, no matter how difficult the road or how heavy your burden. When things seem dark, a smile, even a forced one, will always light the way.

GLOSSARY

Abstract A decorative or mood-setting background that is nonrepresentational and related to no specific location.

Academy leader A numbered strip of film on each reel of a feature that assists projectionists in synchronizing the end of one reel with the beginning of the next. The strip runs for eight seconds and was devised by the Academy of Motion Picture Arts and Sciences.

Account A client or customer of a supplier or advertising agency who purchases goods and services for advertising production.

Account conflict A situation resulting when competing clients—for example, companies manufacturing the same product—are handled by the same supplier.

Account executive The agency employee who maintains liaison between the advertising agency (or production company) and the client; develops and controls the business for the company represented. Also known as contact, account representative, or account manager.

Account supervisor Individual who supervises the work of the account executive(s).

Action 1. Movement of a subject before the camera. 2. A director's signal for such movement and the shooting of that movement.

Agent An individual who is paid to negotiate the buying or selling of goods and services without taking title to such goods or services.

Air date A position or broadcast date for a television or radio program or commercial.

Alternative sponsorship Occurs when two or more companies each buy a segment of time to advertise in turn—whether day by day, week by week, or program by program—within the same television or radio time slot.

Ambient light Undirected light that provides general illumination for a scene or set.

Ambient sound Sound reflected from interior sources rather than picked up directly from a sound source.

Angle shot A shot continuing the action of a preceding shot, but from a different angle.

Animation The creation of the impression of movement.

Answer print The first print of a film in release form that combines sound and picture (if a sound film). It is offered by the lab to the producer and/or director for approval prior to ordering quantity prints.

Apple box A square riser used to raise the height of a performer or prop.

Art director 1. In film or television, the designer and supervisor of set construction. 2. In an advertising agency, the person responsible for the development of design and the supervision of final artwork for ads.

Baby spotlight A small spotlight, often used on faces. Also known as dinky, inkie, inky, or inky dink.

Background The setting over which animated cels are photographed. In live-action film, background is the setting behind the action.

Background light The light that illuminates the shot's background.

Backlight A light behind actors or objects on the set that renders increased separation from set background and gives the impression of depth.

Back lot The portion of a major studio equipped with streets, false-front buildings, and other details that simulate location shooting.

Barndoors Opaque (usually matte black) metal sheets hinged to a frame that are placed in front of a light to permit control of the light.

Barney Lightweight, padded covering placed around a camera to insulate motor sound. Sometimes a barney is heat-wired to assist camera function during extreme cold.

Beat A predesignated pause in an actor's delivery.

Bit A single action or scene.

Blimp A covering for a camera, usually hard and shaped exactly to the camera and its fittings; used in the same manner as a barney.

Bounce light Indirect lighting, usually reflected off walls, ceilings, or reflectors.

Broad A 2,000-watt light in a boxlike lamphouse; used as a soft floodlight for illuminating wide areas.

Burn-in A photographic double exposure; usually used for titles.

Buyout A one-time payment to talent for all rights to a performance (as opposed to a residual schedule).

Camera angle The camera's viewpoint.

Canvass A round of visits or phone calls to new or regular clients to increase sales or business.

CATV Community antenna television; a subscription service wherein individual homes are wired from a central antenna.

CCTV Closed circuit television; used for internal communication in large institutions.

Cel A transparent sheet of plastic on which a figure to be animated is painted.

Client Individual or business who employs the services of an advertising agency or production house.

Clio Award presented at the American Television and Radio Commercials Festival.

Close-up (CU) Emphasis shot that calls attention to a face, inscription on an object, or any other person or thing viewed at close range.

Color correction Modification of tonal values with the use of tinted filters; usually done at dry-lab printer.

Color temperature In photography, the degrees Kelvin (K) of a light source. High temperatures produce blue tones or cast; low temperatures produce red tones.

Composite A synchronization of picture and sound in a short piece of film. See also *interlock*.

Composition In photography, the relative relationships between objects and backgrounds before the camera; also, a consideration in light and shade balance.

Consumer advertising Advertising directed at the public as a whole rather than at a specific profession or industry.

Contingencies Budget allowances for unforeseen circumstances that might delay the shoot schedule and increase expenses.

Continuity The progressive flow of events in the story line showing logical development of characters and plot.

Continuity cutting Editing film or tape to present action in a smooth, logical flow that preserves the illusion of reality for the audience.

Cookie A patterned screen placed in front of a light to form interesting patterns on the set background.

Copy Writing for advertising.

Copy chief Formerly, the agency supervisor of writing; now archaic since art and copy departments have merged.

Copywriter Individual who writes advertising or editorial copy.

Corporation for Public Broadcasting (CPB) Administers federal monies allocated for public broadcasting.

Crawl roll A drumlike mechanism on which titles and credits are placed for filming or taping.

Creative Pertaining to the process of conceiving, developing, and executing advertising ideas.

Creative director (CD) Individual who manages agency creative personnel, such as writers and artists.

Credits Acknowledgment and identification of actors and crew members at the end of a film or video; also the title of the work at both the opening and closing of a film.

Cucalorus Spotlight screen or filter used to project a specific shape in shadow or outline a form on a backdrop. Also known as cuckoo-lorus, cukaloris, cookie, and cuke.

Cue card A large card bearing the lines to be spoken by a performer. Also known as idiot card.

Cut 1. An instantaneous scene change. 2. Director's signal to stop shot and action. 3. A physical cut on film segments during the editing process.

Cyclorama A curved backdrop used to give the effect of sky or distance. Also known as cyc.

Dailies Work prints made on a daily basis as the shoot continues, from which the best takes are selected. Also known as rushes.

Day for night Daylight exterior shots that simulate night by using filters.

Depth of field The amount of photographed area or object that is in sharp focus compared to foreground and background; the depth of field narrows as the lens aperature opens.

Direct mail Use of the postal system to deliver advertising.

Director The individual who interprets the script and supervises its filming.

Dissolve A superimposition of two scenes—one fading in and the other fading out—varying in length for desired effect.

Documentary Presentation of actual events within a time frame.

Dolly A small, wheeled platform for the camera that is used to shoot movement before the action.

Door opener An inexpensive gift from a salesperson offered as an inducement to gain a prospect's attention.

Double exposure In photography, either a superimposed shot or two or more shots taken at different times on the same film. Also known as side-by-side or vignette.

Dramatic scene A confrontation.

Dubbing Rerecording dialogue, music, or sound effects to complete the sound track of a film.

Editor In video and filmmaking, the individual who cuts the film and organizes the takes to produce a final print. The cutting is done electronically in video.

Educational television (ETV) The transmission of academic instruction for home and classroom viewing.

Effect Technical creation of visual or auditory illusion.

Elements package All negative film elements essential to produce a master from which prints are derived.

Emulsion speed The photographic sensitivity of a film stock to light exposure; usually expressed as an index number.

Envelope stuffer A printed advertising piece enclosed with a bill or other matter in a mailing envelope.

Episodic Script or film structure giving strong emphasis to incidents while de-emphasizing continuity and the progression toward a climax. *Jeremiah Johnson* and *M*A*S*H* are episodic films.

Establishing shot An opening shot providing a comprehensive view of a scene that in subsequent action will be shot from closer positions.

Executive creative director The top person responsible for an agency's creative team.

Exploitation film A low-budget picture for a specialized audience; subjects include gothic, horror, bikers, rock, beach bikinis.

Exposition Introduction of information from the past necessary to the advancement of the plot.

Exposure meter A device that reads the amount of light in a given location.

Exteriors (EXT) Outdoor shots.

Extreme close-up (ECU) A very strong emphasis shot, as when a mouth fills the frame.

Extreme long shot (ELS) A shot that reduces the size of a subject in relation to its background to a greater degree than a long shot.

Fade A gradual obliteration of an image by means of steadily closing down the camera aperature.

Feathered light The creation of shading or falloff by using only the weaker edge of a light.

Fill light A secondary light on a set intended to prevent excessive light and shade contrast from the key light.

Film chain A mechanical and electronic device used to convert the film and/or slides to a video signal.

Film clip Short film footage used for insertion in program material.

Film loop A piece of motion picture film spliced into a continuous and unending sequence.

Film perforation The holes on film used by the sprockets of a camera or projector to advance it.

Filmstrip A motion picture film produced as a series of still photographs to be shown as such.

Fine cut A finished work print, fully edited and ready for approval prior to negative matching and production for distribution.

Fiscal year The period of time designated by an organization as its basic financial planning unit; usually 12 months long, but not necessarily a calendar year.

Flier A printed piece, usually consisting of a single sheet, used for advertising as a handout or mailing piece.

Flip chart A card bearing one of a sequence of messages for use in a presentation.

Floodlamp A light that illuminates a wide area. Also known as flood or scoop.

Focus The degree of clarity of a projected image or as seen through a lens.

Follow focus Focus setting change to keep a specific object in sharp focus as scene is shot.

Frame 1. Individual picture on film or videotape. 2. To compose a shot.

Full shot A shot that takes in all of a subject.

Gaffer An electrical assistant on a production.

Gobbo (Gobo) An opaque screen for cutting down unwanted light from the camera lens.

Golden time Time for which workers are compensated at special overtime rates, as indicated by union contracts.

Graininess On film emulsion, the size and separation of the individual silver halide particles. Very sensitive emulsions are usually grainy.

Grip A general handyman on a production.

Half apple A stand for performers or props; lower than an apple box.

Hook A striking incident, unique action, or the like, used to capture audience interest in the beginning of a picture.

Interlock The edited work print and rough sound track set up in synchronous parallel for approval presentation.

Jump cut A cut that appears abrupt or out of continuity because some element of the action has been omitted, particularly in a shot from the same angle.

Key light The primary light source in a shot.

Kicker A rim light used to define the outline of a subject in order to separate the subject from the background. Also known as edge light, skimmer, and separation light.

Leave behind A document left with a prospect by a salesman at the conclusion of a sales call.

Lip synchronization (Lip sync) The coincidence of lip movement and sounds of speech as normally seen and heard.

Long shot (LS) A shot that relates the subject to the background; often used as an orientation or establishing shot.

Market research Research necessary to supply data for effective marketing of consumer goods and services.

Marketing The process of product or service development, pricing, packaging, advertising, merchandising, sales, and distribution.

Marketing director The individual responsible for the review and approval of marketing plans, sometimes including sales management.

Marketing plan The strategy for marketing a product or service.

Master shot The long shot in which all action in a scene takes place. The same action is repeated for the medium shot (MS) and close-up (CU) that are usually cut into the scene.

Medium close-up (MCU) The shot between a medium shot and a close-up.

Medium long shot (MLS) The shot between a medium shot and a long shot.

Medium shot (MS) A shot of the subject with only incidental background.

Merchandising In the sales promotion process, making goods or services attractive and conspicuous.

MOS An abbreviation denoting silent filming; from "mit out sound," originated by an old German director with a heavy accent.

Narration Voice-over commentary that is spoken off screen by a voice that may or may not be identified.

News peg A newsworthy item to be fitted into a news program, not as a feature but as a single item, point of reference, or news fill.

News release A document of informational material on a recent or current event, as within a business organization, distributed to broadcasting stations, newspapers, and magazines for public relations purposes. Also known as press release.

Optical effects Optical or electronic manipulation of a motion picture scene. Methods include wipes, dissolves, and fades.

Pan A horizontal camera movement to make a panoramic shot.

Pencil test The filmed rough sketches in an animation sequence prepared prior to cleanup and transfer to cels.

Per diem A daily cost allowance or fee for travel expenses or services.

Picture-line standard The standard number of scan lines (horizontal electronic scanning lines) on a television screen. In the United States, 525 lines are the standard.

Piggyback Two unrelated commercials by the same sponsor aired back-to-back; usually paired for purchase as a single unit.

Plot A writer's dramatized plan of action for manipulating audience emotions.

Plot line The story line; also, the line of dialogue essential to the development and/or understanding of the plot.

Postsynchronization Sound-track recording after the picture is completed. Many European films are shot without sound and postsynched in various languages for distribution.

Presynchronization Recording the sound track before filming. This is usually done on animated films.

Process shot A shot in which the foreground action is staged against a backscreen projection of stills or motion picture footage.

Propaganda Communications intended to influence belief and action.

Proposal outline A brief statement of purpose, target audience, concept, and specifications for a proposed screened communication production.

Protagonist A film's central character in whose fate the audience will be most interested and with whom the audience will most identify.

Publicity Information regarding a person, corporation, or product; released for free use by the media.

Public relations Activities of persons or organizations intended to promote good will toward themselves or their goods and services.

Public service advertising Advertising carried by the media without charge to propagate socially important information and to promote good will.

Public Broadcasting Service (PBS) A television system sponsored by private and corporate contributions and public funds such as taxes.

Raw stock Unexposed motion picture film.

Release A legal contract assigning a person's rights to the use of his name, likeness, ideas, or property to another party in return for a stated consideration.

Release print The final print run for distribution after all corrections are done from the answer print.

Residual A royalty paid to a performer or other person by a television or radio station or advertiser for each broadcast of a program or commercial. Rates are usually established by union contract.

Reversal film Film that is processed to produce a positive image instead of a negative. This process is generally used when the clarity of the first-generation print is important.

Rough cut An assembly of all the takes to be used in the work print without fine cutting or trimming and marked effect for transitions. The rough cut is always in sequence per script.

Rushes See *dailies.*

Scene One continuous segment of action in a film.

Scrim A fine, translucent cloth or screen used to diffuse light as a backdrop or over a light.

Segue Smooth transition from one sound to another, particularly in a musical presentation, as from one musical number to another in a medley.

Sequence A series of shots comprising a brief continuity within a film to dramatize a unified thought or theme.

Shot A single run of the camera or the results on film of such a run.

Slide commercial A television commercial with a video sequence made up wholly or in part of slides rather than film.

Sound-effects track The magnetic track that contains the sound effects other than dialogue or music.

Sound track Any magnetic or optical sound recording, whether separate or a mixed composite of music, dialogue, and effects.

Special effects Multiple image, split screen, miniature, or any unusual effect not obtainable without extra manipulative effort.

Storyboard A presentation panel of illustration of the various shots proposed or planned for a television commercial, animated program, or motion picture, with notes regarding filming, audio components, and script arranged in consecutive order.

Story line A screenplay's plot development.

Story treatment A semidramatized, present tense, preliminary structuring of a screenplay.

Stringer News reporter, typically covering the local news in a specific area, working on a part-time basis.

Subplot A story within a story, generally involving subordinate characters and developed in terms of action parallel to that of the main plot. Its purpose is to provide relief from the main plot tension and add interest to the production.

Suspense Uncertainty of outcome; the fear that something will or will not happen.

Swish pan A transitional device in which a pan is so fast that the images blur; often used to change scenes.

Synopsis A brief outline of a proposed film's content.

Take Film or tape of a single shot.

Talent Actors, musicians, and other performers.

Target audience The audience intended to be reached by an advertiser in using a given communications medium or set of media.

Teaser Intriguing pretitle action used to capture audience attention, especially in a television movie.

Title The name of a film or any information inscription on the film.

Trade name A name applied to a type of goods or service furnished by a company that may also have the exclusive character of a trademark; also the name under which a company or person does business. Xerox, for example, is both a trademark and a trade name.

Trade paper or magazine A periodical edited for the interest of persons associated with a specific trade or industry.

Trade show A special temporary exhibit of goods or services for trade buyers, often done in collaboration with other exhibitors. Also known as trade convention.

Transition The smooth passage from one episodic part to another, maintaining audience orientation and the established mood of the film.

Trucking Moving the camera by using a dolly in order to follow the general movement of the subject.

Waist shot A shot of a person from the waist up.

Walk-through An early rehearsal to work out wrinkles for both the performers and the crew. Also known as dry run.

Wild shot A shot with no synchronous sound relationship, such as an establishing shot or a shot used for transition only.

Wild spot A spot commercial announcement for a national or regional advertiser used on local station breaks. Also known as spot announcement.

Wipe An optical transition on film in which a second scene seems to be replacing a first scene in a progressive revelation.

Work print A film print used for final editing to produce the answer print by matching the cuts with the original negative after the work print editing is completed.

Zoom lens A lens with a variable focal length for simulating a camera moving into or out of the scene.

SELECTED BIBLIOGRAPHY

Author's note: Publishers have finally discovered filmmaking, so there is now a plethora of books available. Most of them are a waste of time from the practical view, and they reflect the new industry of intellectual film studies developed by the various colleges and universities. As a filmmaker, film studies might be of some value, but it is not mandatory nor as worthwhile as investing the same amount of time studying the business. Do not confuse film-study books with fan picture books, which are strictly for the idle rich. A film-study book will generally be text heavy and picture light, will attempt to analyze its subject(s), and will at least make an effort to say something worthwhile. A fan book shows pictures with some captions that are just trimming on the cake. The best investment in time and money are books on the actual craft of creating, and written by professionals, not "professors" per se. Some professors have practical experience, but their numbers are few. However, there are a few film-study books that cross over into business, and there are some professors, former professionals, who have written excellent books. A partial list follows.

If you have now read this book, you will correctly suspect that I have sold not one part of my soul in the process of attaining senior status in film. If this publisher had asked me to make the following statement, I would not have made it: "When I started looking for a new publisher for this book I suspected it would not take too long to find one." Indeed, I had four offers out of five inquiries to get a second edition in print. I selected Focal Press because I was impressed with its dedication to the screen and the quality of its books in print. This publisher is far from a Johnny-come-lately in our industry. I strongly recommend that the reader obtain the Focal catalog. It contains so many good books for your acquisition that I have listed none here and leave the task to you.

The following books are recommended reading:

GENERAL

Barnouw, Eric. *DOCUMENTARY—A History of the Non-Fiction Film*. Oxford, England: Oxford University Press, 1983.

Bart, Peter. *FADE OUT—The Calamitous Final Days of MGM*. New York: Doubleday, 1991.

Brouwer, Alexandra, and Thomas Lee Wright. *Working in Hollywood*. New York: Avon Books, 1990.

Goldman, William. *Adventures on the Screen Trade*. New York: Warner Books, 1983.

Kronigsberg, Ira. *The Complete Film Dictionary*. Meridian, 1987.

THE BUSINESS OF FILMMAKING

Curran, Trisha. *Financing Your Film*. New York: Praeger Publishers, 1986.

Goodell, Gregory. *Independent Film Production*. New York: St. Martin's Press, 1982.

Klein, Walter J. *The Sponsored Film*. New York: Hastings House, 1976.

Leedy, David J. *Motion Picture Distribution*. Privately published, 1980 (P.O. Box 27845, Los Angeles, CA 90027).

Mayer, Michael F. *The Film Industries: Practical Business/Legal Problems in Production, Distribution, Exhibition*. New York: Hastings House, 1973.

Robertson, Joseph F. *Distribution Handbook*. Summit, PA: Tab Books, 1981.

Rosen, David, with Peter Hamilton. *Off Hollywood—The Making & Marketing of Independent Films*. New York: Grove Weidenfeld, 1990.

Russo, John. *Making Movies*. New York: Delacorte Press, 1989.

Squire, Jason E., ed. *The Movie Business*. Fireside, 1988.

Wiese, Michael. *The Independent Filmmaker's Guide: How to Finance, Produce, and Distribute Your Short and Documentary Films*. Sausalito, CA: M. Wiese Film Productions, 1981.

SCREENWRITING

Edmonds, Robert. *Scriptwriting for the Audiovisual Media*. New York: Teachers College Press, 1978.

Egri, Lajos. *The Art of Dramatic Writing*. New York: Simon & Schuster, 1960.

———. *The Art of Creative Writing*. New York: Simon & Schuster, 1965.

Giustini, Rolando. *The Film Script: A Writer's Guide*. Englewood Cliffs, N.J.: Prentice Hall, 1980.

Nash, Constance, and Virginia Oakey. *The Screenwriter's Handbook*. New York: Barnes & Noble Books/Harper & Row, 1978.

REFERENCE BOOKS FOR THE FILMMAKER

In the first edition, I listed some books in this section, but the years flow quickly and most of the reference books are issued annually. This time around, here is a list of three major sources for good reference books:

The International Motion Picture and Television Almanacs, now including Video. (published annually) Quigley Publishing Company, Inc., 159 West 53rd Street, New York, NY 10019 (212) 247-3100. (Also check with your local library for these; they are not inexpensive.)

The Hollywood Reporter Blu Book Annual Directory (Essentially West Coast stuff, but very comprehensive.) Blu Book Orders: (213) 464-7411 ext. 250 (major credit cards) 6715 Sunset Boulevard, Hollywood, CA 90028.

Barnes& Noble: One of the largest bookstores in the world, it generally carries all film and video books in current print, including curriculum textbooks. 105 Fifth Avenue & 18th St., New York, NY 10003. Book information: (212) 807-0099. Catalog Sales: 1-800-242-6657 (all major credit cards). Barnes & Noble College Book store: 120 Fifth Avenue, New York, NY 10003, (212) 633-4000.

READINGS FOR THE INDEPENDENT SMALL BUSINESSPERSON

General Business Practice and Philosophy

In one of my early books, *THE BUSINESS OF FILMMAKING,* (Rochester, NY: Kodak, 1978) now out of print, the entire area of general business activities was covered; that is, the structuring of your business, whether to form a partnership or a corporation, licenses, taxes and government regulations, insurance, locating your business (selecting the city), managing your finances, keeping financial records (such as costs and equipment depreciation), maintaining a balance sheet, cost control, warning symptoms of financial trouble (including causes of business failure), and obtaining management assistance.

There is also a checklist:

- Personal inventory—are you the type to work for yourself?
- What are your chances for success?
- What will be your return on investment?
- How much capital will you need?
- Where can you get the money?
- Should you share the ownership of your business with others?
- Where should you locate?
- How will you price your service?
- What are the best methods of selling and promoting your films?
- What other management problems will you face?
- What records should you be prepared to keep?
- What insurance problems will you have?

There is also a glossary of financial terms in the book.

As you can see, being in business for yourself can be complicated, and there is a lot of paperwork involved in order to actually operate a business. Because of this, if you are now operating as an independent or plan to be, learning all you can on how to conduct and manage a business is absolutely mandatory. Many businesses fail not just because of undercapitalization, but because of simple, needless ignorance. Operating a business without understanding the basics is much like climbing onto a large machine, like a backhoe, and operating it without training.

Three professionals you will need are an attorney, an accountant, and a business consultant or manager. You can also advance more quickly by taking the time to learn all about business and how it operates successfully. The short list that follows should help. Keep your eyes open for more business information as you develop your operation:

Ballas, George C. and Dave Hollas. *The Making of an Entrepreneur: Keys to Your Success*. Englewood Cliffs, NJ: Prentice-Hall, 1980.

Baty, Gordon B. *Entrepreneurship: Playing to Win*. Reston, VA: Reston Publishing, 1974.

Book of Business Knowledge. New York: Boardroom Books, 1979.

Brandt, Steven C. *Entrepreneuring: The Ten Commandments for Building a Growth Company*. New York: Addison-Wesley, 1982.

Carrol, Frieda. *Survival Handbook for Small Business*. Atlanta, GA: Bibliotheca Press, 1981.

Cook, Peter D. *Start and Run Your Own Successful Business: An Entrepreneur's Guide*. New York: Beaufort Books, 1982.

Deiner, Royce. *How to Finance a Growing Business*. Woodstock, NY: Beekman Publishers, 1974.

Gordon, Barbara, and Elliot Gordon. *How to Survive in the Freelance Jungle*. New York: Executive Communications, 1980.

Kirzner, Israel M. *Competition and Entrepreneurship*. Chicago: University of Chicago Press, 1978.

Linneman, Robert E. *Turn Yourself On: Goal Planning for Success*. New York: Rosen Press, 1970.

Mancuso, Joseph, ed. *The Entrepreneur's Handbook*. Dedham, MA: Artech House, 1974.

Schollhammer, Hans, and Arthur Kuriloff. *Entrepreneurship and Small Business Management*. New York: John Wiley & Sons, 1979.

Scott, William. *How to Earn More Profits through the People Who Work for You: A Practical Handbook for Managers and Small Business Owners to Hiring, Evaluating and Motivating Employees*. Englewood Cliffs, NJ: Prentice-Hall, 1982.

Stevens, Mark. *How to Run Your Own Business Successfully*. New York: Monarch Press, 1978.

Stickney, John. *Self-Made: Braving an Independent Career in a Corporate Age*. New York: Putnam, 1980.

White, Richard M. *The Entrepreneur's Manual: Business Start-Ups, Spin-Offs, and Innovative Management*. Radnor, PA: Chilton, 1976.

Advertising

Amsteil, Joel. *What You Should Know about Advertising*. Dobbs Ferry, NY: Oceana Publications, 1979.

Anuta, Larry. *The Complete Mail Sales Promotion*. Concord, CA: Surevelation, 1977.

Fletcher, Winston. *Teach Yourself Advertising*. Teach Yourself Series. New York: David McKay, 1978.

Garfunkle, Stanley. *Developing the Advertising Plan: A Practical Guide*. New York: Random House, 1980.

Hodgson, Richard S. *Direct Mail and Mail Order Handbook*. Chicago: Dartnell Corporation, 1980.

Hoke, H. *What You Should Know about Direct Mail*. Dobbs Ferry, NY: Oceana Publications, 1966.

Kleppner, Otto. *Advertising Procedure*. 8th ed. Englewood Cliffs, NJ: Prentice-Hall, 1983.

Kincaid, William M. *Promotion: Products, Services and Ideas*. Columbus, OH: Charles E. Merrill, 1981.

Kuswa, Webster. *Big Paybacks from Small Budget Advertising*. Chicago: Dartnell Corporation, 1982.

Norins, Hanley. *The Compleat Copywriter: A Comprehensive Guide to All Phases of Advertising Communication*. Melbourne, FL: R. E. Krieger, 1980.

Pokress, E. *Advertising and Public Relations*. Allenhurst, NJ: Aurea, n.d.

Smith, Cynthia, S. *How to Get Big Results from a Small Advertising Budget*. New York: Dutton. 1973.

Sutton, Cort. *Advertising Your Way to Success: How to Create Best-Selling Advertisements in All Media*. Englewood Cliffs, NJ: Prentice-Hall, 1981.

Wademan, Victor. *Money-Making Advertising: Guide to Advertising That Sells*. New York: John Wiley & Sons, 1981.

Welch, Ditt T., Jr. *Advertising First Class on Any Budget*. Mesquite, TX: Ide House, 1982.

Selling

Adelman, Conrad. *How to Manage Your Sales Time*. Hillsdale, NJ: IBMS, Inc.

Ades, Leslie J. *Increasing Your Sales Potential*. New York: Harper & Row, 1981.

Aronson, Sam. *Everyone's Guide to Opening Doors by Telephone*. San Mateo, CA: S. Aronson, 1981.

Bender, James F. *How to Sell Well: The Art and Science of Professional Salesmanship*. New York: McGraw-Hill, 1971.

Bobrow, Edwin E. *How to Sell Your Way into Your Own Business*. New York: Sales and Marketing Management Magazine, 1977.

Burstein, Milton B. *What You Should Know about Selling and Salesmanship*. Dobbs Ferry, NY: Oceana Publications, 1969.

Ellman, Edgar. *Recruiting and Selecting Profitable Sales Personnel*. New York: Van Nostrand Reinhold, 1981.

Fellows, Hugh P. *Art and Skill of Talking with People: A New Guide to Personal and Business Success*. Englewood Cliffs, NJ: Prentice-Hall, 1964.

Kinder, Jack, Jr., et al. *Winning Strategies in Selling*. Englewood Cliffs, NJ: Prentice-Hall, 1981.

Peterson, Ken T. *How to Sell Successfully by Phone*. Chicago: Dartnell Corporation, 1975.

Steinberg, Jules. *Customers Don't Bite: Selling with Confidence*. New York: Fairchild, 1970.

Public Relations

Barber, Harry L. *How to Steal a Million Dollars in Free Publicity*. Newport Beach, CA: Newport Publishing, 1982.

Benn, Alec. *The Twenty-Three Most Common Mistakes in Public Relations*. New York: American Management Association. 1982.

Bernays, Edward L. *Public Relations*. Norman, OK: University of Oklahoma Press, 1977.

Bowman, P. and N. Ellis. *A Manual of Public Relations*. New York: State Mutual Bank, 1977.

Culligan, Matthew J., and Dolph Greene. *Getting Back to the Basics in Public Relations and Publicity*. New York: Crown, 1982.

Delacorte, Toni, et al. *How to Get Free Press: A Do It Yourself Guide to Promote Your Interests, Organization, or Business*. San Francisco: Harbor, 1981.

Kadon, John, and Ann Kadon. *Successful Public Relations Techniques*. Scottsdale, AZ: Modern Schools, 1976.

Lewis, H. G., *How to Handle Your Own Public Relations*. Chicago: Nelson-Hall, 1976.

MacDougall, Curtis. *Interpretative Reporting*. New York: Macmillan, 1977.

Samstag, Nicholas, *Persuasion for Profit*. Norman, OK: University of Oklahoma Press, 1957.

Simon, Raymond, ed. *Perspectives in Public Relations*. Norman, OK: University of Oklahoma Press, 1966.

Marketing

Bradway, B. M., and M. A. Frenzel. *Strategic Marketing: A Handbook for Entrepreneurs and Managers*. Reading, MA: Addison-Wesley, 1982.

Breen, George Edward. *Do-It-Yourself Marketing Research*. New York: McGraw-Hill, 1981.

Buen, Victor P., and Carl Heyel, eds. *Handbook of Modern Marketing*. New York: McGraw-Hill, 1970.

Luther, William M. *The Marketing Plan: How to Prepare and Implement It*. New York: American Management Association, 1982.

RESOURCES

This list does not contain "fan" or "film studies" periodicals but is skewed toward the film production and film business professional. Some commentary is offered on the more prominent periodicals.

Advertising Age, 740 Rush Street, Chicago, IL 60611. Published weekly, this is the oldest and largest trade in advertising. Important for TV-spot producers.

AFI Guide to College Courses in Film & Television, American Film Institute, Education Services, 2021 N. Western Ave., Los Angeles, CA 90027.

American Cinematographer, A.S.C. Holding Corp., 1782 N. Orange Drive, Los Angeles, CA 90028. The journal of the American Society of Cinematographers, important for the serious filmmaker, now also covering TV and video production.

AMERICAN FILM AND VIDEO ASSOCIATION, 920 Barnsdale Rd., LaGrange Park, IL 60525-1609. They publish the three following periodicals.

- *AFVA Bulletin,* news tidbits on nontheatrical film and video production.
- *AFVA Evaluations,* brief reviews on documentary and personal 16mm film and video.
- *American Film and Video Festival Guide,* descriptions and distribution information for the 16mm and video producer.

Association of Cinema and Video Laboratories, P.O. Box 34932, Bethesda, MD 20034. *Annual Directory* (updated annually), provides editorial guidelines and standards.

Cinefex, Box 20027, Riverside, CA 92516. Journal of film special effects (optical, physical, computer, makeup).

The Creative Black Books, 115 Fifth Ave., New York, NY 10010. Annual advertising craft showcase, expensive and comprehensive, if not complete.

Directors Guild of America, Directory of Members, Directors Guild of America, Inc., 7950 Sunset Blvd., Hollywood, CA 90046. Annual update on member's list, with some contact information on members that allow listing.

Educators' Guide to Free Films, Educators Progress Service, Inc., 214 Center St., Randolph, WI 53956.

Film Journal (Independent films) Pubson Corp., 224 W. 49th St., 3rd fl., New York, NY 10019. Offers industry background.

Film Producers, Studios, Agents and Casting Directors Guide, Lone Eagle Publishing Co., 9903 Santa Monica Blvd. #204, Beverly Hills, CA 90212. A good source for possible networking contacts. Priced at $30.00 at this writing.

Film Threat, 6646 Hollywood Blvd., #205, Los Angeles, CA 90028. Independent film industry articles and news.

Foundation for Independent Video and Film, Inc., and The Association for Independent Video and Filmmakers (FIVF/AIVF). (See Chapter 6 for more information on these two associated organizations.)

- Bowser, Kathryn. *AIVF Guide to Film and Video Distributors*
- *Independent Film and Video Monthly* (magazine)
- Warshawski. *The Next Step: Distributing Independent Films and Videos*
- Franco. *Alternative Visions: Doing it Yourself*

(Also check this organization for various additional publications.)

Hollywood Production Manual, 1322 North Cole Avenue, Hollywood, CA 90028. Annual, with periodic updates. This is the filmmaker's bible in the feature and commercial areas, offering current rates, budget items, and other production problem forecasts. Not cheap, folks; this is for real players.

Hollywood Reporter, 6715 Sunset Blvd., Hollywood, CA 90028. Weekly industry news, also *Blu Book, Annual Directory*, A good contact and networking source, it has gotten better and more comprehensive with each passing year. This directory is roughly $55.00 (See Chapter 10 for other information.)

In Motion Magazine, 1203 West Street, Annapolis, MD 21401. Professional journal.

Independent, Foundation for Independent Video and Film, 625 Broadway, New York, NY 10012. The trade for independent video and filmmakers.

Index to Kodak Information, 343 State Street, Rochester, NY 14650. Annual Guide to the technical literature published by Kodak. Ask for pamphlet L-5. Worthwhile, comprehensive, and free.

International Motion Picture Almanac and *International Television Almanac*, Quigley Publishing Co., 159 W. 53rd St., New York, NY 10019. Feature and TV directories; expensive, usually available at the larger libraries. Includes a Who's Who in the industry.

Legal and Business Problems of Financing Motion Pictures, Practicing Law Institute, 810 Seventh Ave., New York, NY 10019.

Millimeter Magazine, 826 Broadway, New York, NY 10003. Far less important to the producer; advertiser-heavy, highly video-technical. This once important magazine is now legible only to the technicians of the industry with little to offer the producer-director. Listed here as a source for video hardware information.

Motion Picture Investor, Paul Kagan Assocs. Inc., 126 Clock Tower Pl., Carmel, CA 93923. Data on motion picture industry, private and public values of movies and movie stock. Very expensive, for the heavy players and investors only.

Off-Hollywood Report Independent Feature Project, 132 W. 21st St., New York, NY 10011. This organization has been helpful to independent filmmakers.

Producer's Masterguide by Shmuel Bension, Publisher and Editor-in-Chief (The International Production Manual for Motion Picture, Television, Commercials, Cable, and Videotape industries), 330 West 42nd Street, 16th flr., New York, NY 10037. (212) 465-8889; fax: (212) 465-8880.

Considered the single most indispensable reference book in the industry by everyone who has ever used it, both for newcomers and seasoned professionals. $98.95; abroad: U.S. $128.95 (12th edition at this writing) and truly priceless. A jewel.

Producer's Quarterly, 25 Willowdale Ave., Port Washington, NY 11050. Very good, comprehensive, industry-wide professional's journal covering the spectrum right through to commercial and corporate TV/film.

SMPTE Journal, Society of Motion Picture Technicians and Engineers. The technician's journal. A good one.

Sightlines, American Film and Video Association, 920 Barnsdale Rd., LaGrange Park, IL 60525-1609. 16mm and video production news.

ORGANIZATIONS THAT AID THE FILMMAKER

Telephone and fax numbers are included where deemed essential.

American Film Institute
The John Kennedy Center for the Performing Arts
Washington, DC 20566

American Society of Cinematographers
P.O. Box 2230
Hollywood, CA 90078
(213) 859-7730
fax (213) 876-4973

American Marketing Association
250 South Whacker Drive
Chicago, IL 60606-5819

American Museum of the Moving Image
34-12 36th St.
Astoria, NY 11106

Arbitron
142 West 57th St.
New York, NY 10019

West Coast office
3333 Wilshire Blvd.
Los Angeles, CA 90010

Association of Independent Commercial Producers (AICP)
100 E. 42nd St.
New York, NY 10017
(212) 867-5720

National office
34-12 36th St.
Astoria, NY 11106
(718) 392-2427
West Coast office
2121 Avenue of the Stars, Suite 2700
Los Angeles, CA 90067-5010
(213) 557-2900

Association of Independent Video and Filmmakers
625 Broadway
New York, NY 10012
(212) 473-3400

Directors Guild of America, Inc. (DGA)
7920 Sunset Blvd.
Los Angeles, CA 90046
(213) 289-2000
fax: (213) 289-2029
East Coast office
110 West 57th St.
New York, NY 10019
(212) 581-0307

(DGA) Producer's Pension & Welfare Plans
8436 W. Third St., Suite 900
Los Angeles, CA 90048
(213) 653-2991

National Captioning Institute, Inc.
5203 Leesburg Pike
Falls Church, VA 22041

Writers Guild of America, East, Inc.
555 West 57th St.
New York, NY 10019
(212) 767-7800

Writers Guild of America, West, Inc.
8955 Beverly Blvd.
Los Angeles, CA 90048
(213) 550-1000

The Producer's Masterguide, the *Hollywood Reporter Blu Book* and other reference books listed in the preceding periodical section carry a complete listing of craft and technical unions and other organizations.

INDEX